Engineering Materials

This series provides topical information on innovative, structural and functional materials and composites with applications in optical, electrical, mechanical, civil, aeronautical, medical, bio- and nano-engineering. The individual volumes are complete, comprehensive monographs covering the structure, properties, manufacturing process and applications of these materials. This multidisciplinary series is devoted to professionals, students and all those interested in the latest developments in the Materials Science field, that look for a carefully selected collection of high quality review articles on their respective field of expertise.

More information about this series at http://www.springer.com/series/4288

David May

Integrated Product Development with Fiber-Reinforced Polymers

Springer

David May ⓘ
Leibniz-Institut für Verbundwerkstoffe
GmbH
Kaiserslautern, Rheinland-Pfalz, Germany

Translation from the German language edition: *Integrierte Produktentwicklung mit Faser-Kunststoff-Verbunden* by David May, © The Editor(s) (if applicable) and The Author(s), under exclusive license to Springer Nature Switzerland AG 2021. Published by Springer Nature. All Rights Reserved.

ISSN 1612-1317   ISSN 1868-1212 (electronic)
Engineering Materials
ISBN 978-3-030-73409-1   ISBN 978-3-030-73407-7 (eBook)
https://doi.org/10.1007/978-3-030-73407-7

This Springer imprint is published by the registered company Springer Nature Switzerland AG
The registered company address is: Gewerbestrasse 11, 6330 Cham, Switzerland

Preface

Fiber-reinforced polymers (FRPs) offer outstanding advantages, with the high lightweight potential being the most striking but by far not the only one. The industrial relevance of FRP is therefore steadily increasing, but still doesn't match the outstanding potential of this fascinating group of materials. One of the main reasons for this is given by the lack of dissemination of FRP-related expertise. To overcome this obstacle, it is necessary to educate future generations of engineers accordingly. This again requires product developers, who on the one hand are specialists in their respective fields, but on the other hand are holistically thinking generalists. This book intends to provide future engineers with this way of thinking and the corresponding, necessary skills.

The book starts with an introduction on the material-specific advantages of FRPs and the typical areas of application. Subsequently, it is shown which problems a conventional, non-integrating product development can cause and how integrated product development (IPD) allows to overcome them. In addition, it is explained why IPD is of particular importance for FRPs. In the following chapter, an approach for the IPD with FRPs is presented. The further structure of the book corresponds to this approach and explains it step by step: At first, a requirements catalog is created and then based on this, a concept, a draft, and finally an elaborated design are developed. The approach contains simple but effective methods for the selection of fiber materials, semi-finished products and manufacturing processes. With these methods, fundamental decisions can be made at an early stage of the development process, thereby enabling simultaneous engineering concerning design-, manufacturing- and material-related aspects.

A concluding chapter describes an approach to techno-economic evaluation, allowing an engineer to decide which of the different developed alternatives to follow up on. Hence, the book covers the entire product development from the definition of the task to the final decision-making. Application examples show how the acquired knowledge can be put into practice. At the end of each main chapter, questions on the covered topics allow self-checking of the learning progress.

For didactic reasons, the order of book chapters corresponds to the suggested approach and is not divided into a basic part and an application part. This will facilitate an understanding for readers who are new to the topic.

This book is based on a lecture, which the author gives at the Technical University of Kaiserslautern. Special thanks are given to Prof. Dr.-Ing. Ulf Paul Breuer, Prof. Dr.-Ing. Joachim Hausmann, Prof. Dr.-Ing. Peter Mitschang, Dipl.-Ing. Dr.mont. Harald Grössing, Ass. Prof. Dipl.-Ing. Dr.mont. Ewald Fauster, Dipl.-Wirtsch.-Ing. Claus Becker and Dr. rer. nat. Martin Gurka for the critical, precise and helpful proof-reading.

Kaiserslautern, Germany   David May
2021

Contents

About the Author

David May received his doctorate degree in 2015 from the Technical University Kaiserslautern, where he since then lectures "integrated product development with composites". In 2021 he completed his habilitation on process engineering for composites. He currently manages an interdisciplinary research group at the Leibniz-Institut für Verbundwerkstoffe in Kaiserslautern. Manufacturing technologies for polymer composites are his main research focus.

Abbreviations

1D	One-dimensional
2D	Two-dimensional
3D	Three-dimensional
AF	Aramid fibers
AHP	Analytical hierarchy process
Au	Autoclave
BMC	Bulk molding compound
BVD	Barely visible defect
CAI	Compression after impact
CAPRI	Controlled atmospheric pressure resin infusion
CC	Centrifugal casting
CF	Carbon fiber
CFRP	Carbon fiber-reinforced polymer
C-Glass	Glass with improved chemical resistance
CLT	Classical laminate theory
CM	Compression molding
CNC	Computerized numerical control
C_{seq}	Sequential, complementary process combination
C_{sim}	Simultaneous, complementary process combination
CTE	Coefficient of thermal expansion
DIN	Deutsches Institut für Normung (German institute for standardization)
DSC	Differential scanning calorimetry
DTMA	Dynamic thermo-mechanical analysis
E-Glass	Glass with high electric insulation capability
EN	European standard
EoL	End of life
EP	Epoxy resin
Ex	Extrusion
FE	Finite element

FMEA	Failure mode and effects analysis
FRP	Fiber-reinforced polymer
FS	Fiber spraying
GF	Glass fiber
GFRP	Glass fiber-reinforced polymer
GMT	Glass mat-reinforced thermoplastic
HDT-A	Heat deflection temperature following standard DIN EN ISO 75-1,-2,-3, method A
HL	Hand lay-up
HM	High-modulus carbon fiber
HT	High-tenacity carbon fiber
ILSS	Interlaminar shear strength
IM	Intermediate modulus carbon fiber
InM	Injection molding
IPD	Integrated product development
ISO	International Organization for Standardization
IVW	Leibniz-Institut für Verbundwerkstoffe GmbH
LCM	Liquid composite molding
LFT	Long fiber-reinforced thermoplastic
MIR	Material input rate
NF	Natural fiber
NVD	Non-visible defect
OT	Orthotropic fiber orientation
PA6	Polyamide 6
PA66	Polyamide 66
PBCM	Process-based cost modeling
PCM	Prepreg compression molding
PD	Product development
PEEK	Polyetheretherketone
PEI	Polyetherimide
PES	Polyethersulfone
PP	Polypropylene
PPS	Polyphenylsulfide
Pu	Pultrusion
QI	Quasi-isotropic fiber orientation
rCF	Recycled carbon fiber
RTM	Resin transfer molding
S	Substituting process combination
SCRIMP	Seemann composites resin infusion molding process
SF	Steel fiber
S-Glass	Glass with improved strength
SiC	Silicon carbide
SMC	Sheet molding compound
T	Temperature
TF	Thermoforming

TGA	Thermogravimetric analysis
Ti	Titanium
TL	Tape laying
TP	Thermoplastic
TS	Thermoset
UD	Unidirectional fiber orientation
UHM	Ultra-high modulus carbon fiber
UHMW-PE	Ultra-high-molecular-weight-polyethylene
UP	Unsaturated polyester resin
VD	Visible defect
VDI	Verein Deutscher Ingenieure e.V.
VE	Vinyl ester resin
VI	Vacuum infusion
Wi	Winding
WPC	Wood polymer composite

Symbols (Latin Letters)

∇p	Pressure gradient (Pa/m)
A	Cross-sectional area/contact area (mm^2)
A_{F1}	Cross-sectional area of fiber (m^2)
A_{F2}	Sheath area of fiber (m^2)
A_{NW}	Number of workers (-)
A_j	Required space for production unit j (m^2)
c	Costs per unit (€)
c_W	Costs for production waste (€)
c_E	Energy costs (€)
c_f	Fixed costs (€)
C_{tot}	Total costs (€)
C_i	Total costs for the process i (€)
c_M	Material costs (€)
c_L	Labor costs (€)
c_O	Other operating costs (€)
c_v	Variable costs (€)
cr_W	Cost rate for production waste (€/kg)
cr_E	Cost rate for energy (€/kwh)
cr_{Ml}	Material price for material l (€/kg)
cr_O	Cost rate for other costs (€/h)
cr_{L_s}	Labor cost rate for worker s (€/h)
cr_M	Cost rate for maintenance (%)
cr_S	Cost rate for space (€/m^2)
d_F	Fiber diameter (m)
D_j	Depreciation for production unit j (€)
E	Stiffness (GPa)
$E_\parallel$	Stiffness of a unidirectional fiber-reinforced polymer in fiber direction (GPa)

$E_\perp$	Stiffness of a unidirectional fiber-reinforced polymer transverse to fiber direction (GPa)
$E_{F,\parallel}$	Fiber stiffness in fiber direction (GPa)
$E_{F,\perp}$	Fiber stiffness transverse to fiber direction (GPa)
$E_{Fiber,\parallel}$	Fiber stiffness in fiber direction (GPa)
E_F	Fiber stiffness (GPa)
E_M	Matrix stiffness (GPa)
$E_{Composite,\parallel}$	Stiffness of a unidirectional fiber-reinforced polymer in fiber direction (GPa)
$E_{Composite,\perp}$	Stiffness of a unidirectional fiber-reinforced polymer transverse to fiber direction (GPa)
E_x	Stiffness in x-direction (GPa)
E_y	Stiffness in y-direction (GPa)
F	Force (N)
$F_{1\rightarrow2}$	View factor (-)
FVC	Fiber volume content (%)
FVC_{crit}	Critical fiber volume content (%)
FVC_{min}	Minimum fiber volume content (%)
F_{max}	Maximum allowable force (N)
G	Grashof number (-)
g	Gravitational acceleration (m/s^2)
$G_{\perp\parallel}$	Shear stiffness (GPa)
$G_{F\perp\parallel}$	Fiber shear stiffness (GPa)
G_M	Matrix shear stiffness (GPa)
h	Height (m)
H	Distance between radiation source and receiver (m)
i	Calculative interest rate (%)
I	Second moment of area (m^4)
IC_j	Interest costs for capital bound by production unit j (€)
In_j	Investment costs for production unit j (€)
K	Permeability (m^2)
l	Length (m)
L	Characteristic length (m)
L_1	Distance between bearing 1 and bearing 2 of thread lever (m)
L_2	Distance between bearing 2 and thread eyelet (m)
l_{crit}	Critical fiber length (m)
m_{Cl}	Required quantity of material l per component (kg)
m_P	Frequency of process step P (-)
m_k	Frequency of activity k (-)
m_l	Frequency of element l (-)
MC_j	Maintenance costs for production unit j (€)
n_j	Number of usage periods for production unit j (-)
N	Nusselt number (-)

p	Annual production (parts per year)
P	Prandtl number (-)
Q	Volume flow rate (m^2/s)
$\dot{Q}$	Heat flow (W)
$\dot{q}$	Heat flux density (W/m^2)
q_d	Production downtime rate (%)
q_{Rej}	Production rejection rate (%)
q_W	Production scrap rate (%)
q_{Rec}	Recycling rate (%)
Re	Reynolds number (-)
$R_{Fiber,\parallel}$	Fiber strength in fiber direction (MPa)
R_{Matrix}	Matrix strength (MPa)
RV_{jn}	Residual value of production unit j at the end of the usage phase (€)
RV_{jn-1}	Residual value of production unit j one year before the end of the usage phase (€)
$R_{m,F}$	Fiber tensile strength (MPa)
$R_{m,M}$	Matrix tensile strength (MPa)
s	Thickness (m)
SC_j	Space costs for production unit j (€)
t_{eff}	Effective cycle time (h)
T_1	Temperature on upper side of tool (K)
T_2	Temperature on bottom side of tool (K)
$T_{D,aTP}$	Decomposition temperature of amorphous thermoplastics (K)
$T_{D,TS}$	Decomposition temperature of thermosets (K)
$T_{D,scTP}$	Decomposition temperature of semi-crystalline thermoplastics (K)
T_F	Ambient temperature (K)
$T_{F,aTP}$	Flow temperature of amorphous thermoplastics (K)
$T_{F,scTP}$	Flow temperature of semi-crystalline thermoplastics (K)
$T_{G,aTP}$	Glass transition temperature of amorphous thermoplastics (K)
$T_{G,TS}$	Glass transition temperature of thermosets (K)
$T_{G,scTP}$	Glass transition temperature of semi-crystalline thermoplastics (K)
T_K	Temperature of object K (K)
T_{max}	Maximum operating temperature (K)
$T_{M,scTP}$	Melt temperature of semi-crystalline thermoplastics (K)
T_{Melt}	Melting temperature (K)
T_{Resin}	Resin temperature (K)
T_{Tool}	Tool temperature (K)
$u_{F,max}$	Maximum allowable bending deformation at thread lever (μm)
u_F	Bending deformation at thread lever (μm)
$u_{x,max}$	Maximum allowable elongation in x-direction (μm)
u_x	Elongation in x-direction (μm)
$\vec{v}$	Volume-averaged flow velocity (m/s)

V_F	Fiber volume content (%)
V_{Pkl}	Value of the element l of the activity k, of the process step P (€)
W_A	Energy consumption (kWh)
x	Length variable (m)
z	Edge length (m)

Letters (Greek Symbols)

α	Convective heat transfer coefficient (W/(m^2 K))
β	Thermal expansion coefficient (1/K)
γ	Angle between seat and wall in the example for the complete requirements catalog (°)
ΔL	Flow length (m)
Δp	Pressure drop (Pa)
$\varepsilon_{Fiber,Break}$	Elongation at break of fiber (%)
$\varepsilon_{Matrix,Break}$	Elongation at break of matrix (%)
η	Fluid viscosity (Pa s)
λ_w	Thermal conductivity (W/(m K))
$\nu_{\parallel\perp}$	Transverse contraction of a unidirectionally reinforced polymer in fiber direction under load transverse to the fiber direction (-)
$\nu_{\perp\parallel}$	Transverse contraction of a unidirectionally reinforced polymer transverse to the fiber direction under load in fiber direction (-)
$\nu_{F\parallel\perp}$	Fiber contraction in fiber direction under load transverse to fiber direction (-)
$\nu_{F\perp\parallel}$	Fiber contraction transverse to fiber direction under load in fiber direction (-)
ν_M	Transverse contraction number of the matrix (-)
π	Ratio of a circle's circumference to its diameter (-)
ρ_{Fiber}	Fiber density (kg/m^3)
ρ_{Matrix}	Matrix density (kg/m^3)
σ	Stefan-Boltzmann constant (W/(m^{2}K^4))
σ'_m	Stress at matrix when fiber breaks (MPa)
$\sigma_{F,Break}$	Breaking stress of fiber (MPa)
σ_F	Stress on fiber (MPa)
τ	Shear stress (MPa)
τ_{max}	Interface shear strength (MPa)

υ   Kinematic viscosity of air at 500 °C (Pa s)
φ   Fiber volume content (%)
ψ   Fiber mass content (%)

Chapter 1
Introduction

Abstract In this section, the term fiber-reinforced polymer (FRP) is defined. Subsequently, the industrial importance of FRPs is shown and specific characteristics of FRPs, which are the reason for this importance, are explained. Afterward, the differences between conventional and integrated product development are presented, and it is shown why integrated product development is of such great importance, especially for FRPs.

Keywords Product Development · Fiber-reinforced polymers · Fiber paradoxon

1.1 Working Principles in Fiber-Reinforced Polymers (FRPs)

Fiber-reinforced polymers are classified as composite materials or simply composites. Composites are defined as materials that result from the combination of two or more materials, whereas the composite shows properties that are at least partially superior to those of the individual material partners. In FRPs, one (or more) of the material partners is present in the form of fiber, while another material partner is a polymer that forms the matrix and surrounds the fibers (Fig. 1.1).

This book focuses on FRPs, whereby with regard to polymers thermosets, thermoplastics as well as certain hybrid forms are considered. Fiber-reinforced elastomers, however, are not considered, as they strongly differ in their characteristics and application areas. The same applies to other fiber composite materials, such as fiber-reinforced metals or ceramics, and other composite structures (e.g. particle-reinforced materials).

The high lightweight design potential of FRPs mainly results from the advantages of the fiber form. This principle was formulated in the 1920s by Griffith [1]: "A material in fiber form has a much higher strength in the direction of the fibers than the same material in another form. The thinner the fiber is, the greater is its strength." This is illustrated in Fig. 1.2.

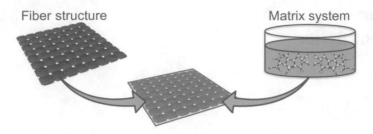

Fig. 1.1 Basic structure of fiber-reinforced composites

Fig. 1.2 Influence of the fiber diameter on the tensile strength of fibers. Adapted from [2]

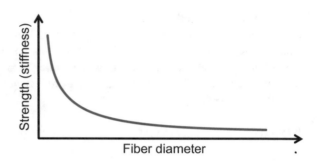

Common fiber materials, such as glass or carbon fibers, make use of this effect. Their diameter is in the micrometer range. Four causes can be identified to explain this effect [3]:

Size effect: The diameter of a fiber limits the maximum possible size of a defect. At an equal total volume, many thin fibers, compared to many thick fibers, statistically show smaller defects, which has a positive effect on the strength. This can be illustrated well with glass, a relatively brittle material. Under tensile load, a tiny defect in a block of bulk glass can quickly propagate as a crack through the entire block. In a bundle of fibers, the same defect would only lead to failure of a single fiber. Given an equal overall cross section, the fibers in the bundle therefore exhibit greater strength.

Molecular alignment: When fibers are produced under tensile load (referred to as drawing), the strongest atomic bond is oriented in the longitudinal direction of the fiber, which means that the fiber becomes mechanically anisotropic with improved mechanical properties in the longitudinal direction of the fibers. This effect occurs in materials with a crystalline structure, such as carbon fibers in which the graphite planes are aligned, but also in polymer fibers, where the molecular chains align. However, such an effect is not possible with all materials, e.g. not with glass, due to its amorphous structure.

Defect orientation: The drawing of a fiber causes any existing defects to be oriented in the longitudinal direction of the fiber, so that their notch effect is reduced. This is a significant effect, especially for brittle materials such as glass.

Residual stresses: Especially with glass fibers, the fiber pullout from the melt causes faster cooling on the outside compared to the inside. This creates thermally induced residual compressive stresses, which reduce the strength-reducing effect of surface damages.

The fiber form is thus responsible for the advantageous mechanical properties of FRPs, but this potential can only be exploited through the special **distribution of tasks**, in combination with the polymer matrix:

The **fibers** attract the forces due to their comparatively high stiffness.

The **matrix**

- embeds the fibers,
- fixates them,
- supports them under pressure,
- thus enables the load to be introduced into the fibers,
- thereby provides load distribution between the fibers and
- protects the fibers from environmental influences.

Figure 1.3 illustrates this distribution of tasks with regard to load introduction. In principle, loads are applied to the matrix and are passed on to the fibers by shear. Under pressure, the matrix also prevents buckling of the fibers. The load transfer between fiber and matrix takes place in the so-called **interphase**. This phase determines the adhesion between fiber and matrix. Formerly, the interphase was often modeled as a pure shell structure without depth, but today FRPs are commonly described as a three-phase model (Fig. 1.4), since the bonding agents applied to the fibers form an area showing different properties compared to the rest of the

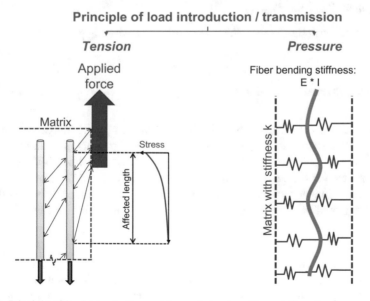

Fig. 1.3 Principle of load introduction and transmission in a fiber-reinforced polymer

Fig. 1.4 FRP as three-phase model

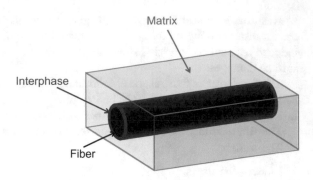

matrix [4]. Depending on the material combination, the interphase can have a thickness of <100–300 nm [5]. This means that depending on the fiber diameter and the fiber volume content, the interphase can represent 20% or more of the entire matrix.

In order for this distribution of tasks to emerge and for a FRP with the desired advantageous properties to result, various **effectiveness criteria** must be met [6]:

1. The Young's modulus of the fibers in the longitudinal direction must be higher than that of the matrix, so that the fibers bear the majority of the loads: $E_{F,\parallel} > E_M$.
2. The elongation at break of the matrix must be higher than that of the fibers so that the fiber strength is exploited: $\varepsilon_{Matrix,Break} > \varepsilon_{Fiber,Break}$.
3. The tensile strength of the fibers in the longitudinal direction must be greater than that of the matrix, so that there is no weakening compared to the pure polymer: $R_{Fiber,\parallel} > R_{Matrix}$.

If all these criteria are met, there are various possible applications.

1.2   Industrial Use of FRPs

In 2018, the German Federation of Reinforced Plastics (AVK Industrievereinigung verstärkte Kunststoffe e. V.) estimated a European production volume of glass fiber-reinforced polymers (GFRP) of about 1,141[1] kt (Fig. 1.5). GFRPs therefore represents by far the largest used reinforcement among FRPs. The significantly smaller, but stronger growing category of carbon fiber-reinforced polymers (CFRP) showed a global annual demand of 128 kt in 2017 (as estimated by the Carbon

---

[1]This number does not include short fiber-reinforced thermoplastics, with a fiber length smaller than 2 mm, which are mainly processed by injection molding. In 2017, the European demand for these materials only exceeded 1.400 kt [7].

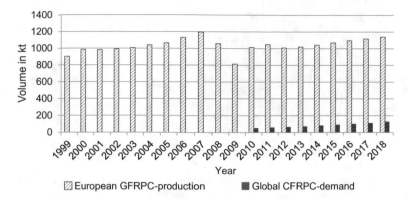

Fig. 1.5 European production volume of GFRPs and global demand for CFRPs. Data from [7], numbers for 2018 are estimated

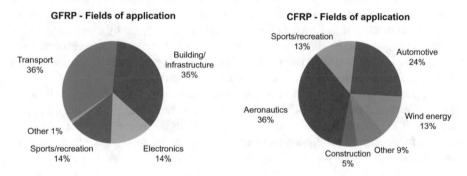

Fig. 1.6 Fields of application for GFRPs and CFRPs. Data from [7]

Composites e. V.). In addition, there are other FRPs such as natural or polymer fiber-reinforced polymers [7].

As shown in Fig. 1.6, FRPs are used in a variety of industries. Transportation and construction are the largest areas of application for GFRPs. For CFRPs, which are classically considered as high-performance materials, the aeronautics/space industry clearly provides the main area of application. However, they can also be found in sporting goods, and in the fields of mechanical and automotive engineering [8].

FRPs as a material group are extremely diverse and, depending on their design, FRPs can offer a whole range of specific advantages compared to other materials:

Lightweight potential: The most outstanding property of FRPs is their high specific (mass density related) stiffness and strength. Figure 1.7 (left) shows a comparison with other materials in terms of the specific properties. Here, conventional steel is used as a reference, to which the value 1 is assigned. It can be seen that FRPs with quasi-isotropic behavior (no preferred orientation of the fibers; marked in the diagram by triangles) already provide double to triple the specific

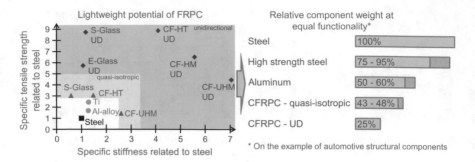

Fig. 1.7 Lightweight potential of FRPs compared to metals. Adapted from [9] (Image adapted, printed with permission of Roland Berger Holding GmbH)

stiffness/strength of steel. They also surpass typical metallic lightweight materials, such as aluminum alloys and titanium (marked in the diagram by circles). If the possibility to orient the fibers according to the load direction is fully exploited (UD = unidirectional fiber orientation, marked in the diagram by diamonds), up to seven times the specific stiffness and up to nine times the specific strength can even be achieved. Thus, a component made of FRP can be significantly lighter compared to one made of steel, at equal (load-bearing) functionality.

Obviously, this representation is not universally valid. Nevertheless, it impressively illustrates the lightweight potential that can be achieved in suitable applications. This potential can be exploited, for example, in the transport sector to reduce fuel/energy consumption and emissions and to increase payloads. For sports or medical equipment, e.g. handling can be improved. In mechanical engineering, lightweight design can increase performance. Besides the excellent specific properties, the lightweight design potential of FRPs is further increased by the high geometric flexibility.

Tailored deformation behavior: The possibility to adapt the fiber orientation to the local load situation is not only of interest in terms of lightweight design. A targeted anisotropy, i.e. directional dependence in the properties, also makes it possible to variably adjust stiffness and flexibility in a component. For example, in the case of bicycles made of CFRP, a specifically adjusted flexibility in the direction of travel can increase comfort, while simultaneously a high lateral stability can be achieved [10]. Compared to isotropic materials, the designer has additional degrees of freedom, resulting from the fiber orientation. Figure 1.8 shows another example. The "Flexshaft" is a torsion shaft, which is used in wind turbines to enable the power transmission from the rotor blades to the generator. In larger facilities, for example in the offshore area, this requires shafts that are several meters long. The Flexshaft, which is made of CFRP, has a fiber orientation that makes it comparatively flexible, which, in case of axial displacement of rotor and generator, reduces the bearing loads and thus costly maintenance work can be reduced. On the other hand, torsional stiffness is extremely high and at a length of 8.6 m and a weight of 4.6 tons torsional moments up to 5000 kNm are transmittable [11, 12].

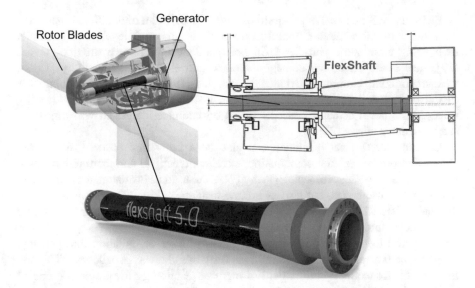

Fig. 1.8 Flexshaft for load transmission in wind turbines. Printed with permission of SchäferRolls GmbH & Co. KG

The possibilities of the anisotropic design also include the targeted adjustment of deformation couplings due to an asymmetrical fiber distribution or—orientation. For example, a FRP can be designed in a way that it will bend or twist when a tensile load is applied. This can, for example, be used to reach a specifically tailored deformation behavior when aerodynamic loads act on the wings of an aircraft [13].

Wide range of materials and manufacturing processes: FRPs consist of at least two material partners, which already multiplies the possible combinations. At the same time, the polymers alone already offer an extreme variety of variants, which is further increased by possible chemical modifications and fillers, such as ceramic particles, carbon nanotubes and metallic powder or elastomer particles. Additionally, there is a high degree of diversity with regard to fiber reinforcement, not only concerning the materials, but also the structure of the fiber reinforcement. This material variety is also reflected in the manufacturing processes. Thus, manufacturing processes suitable for large series production as well as for single-part and small-batch production are available. Some of the processes allow the manufacturing of geometrically highly complex, integral structures.

Functional integration: The geometric design freedom resulting from the variety of manufacturing processes also leads to a high potential for functional integration [14, 15], which in turn can contribute to weight and cost savings. Especially the integration of load-bearing functions can be carried out, i.e. a single component can take over several structural tasks within a system. This way, the number of components can be reduced, e.g. compared to a conventional steel construction.

FRPs also offer extensive possibilities for the integration of additional materials, with the aim of achieving multifunctionality at material level. For example, toughening elastomeric particles [16], fire-retardant aluminum hydroxide-powder [17], steel fibers increasing damage-tolerance and electrical conductivity [18] or wear-minimizing aluminum oxide nanoparticles [19] can be integrated into the FRP. Even larger materials can be incorporated, such as laminate-integrated heating foils [20] or novel actuator concepts based on laminate-integrated wires made of shape memory alloys [21].

Electrical or thermal insulation can also be a functional property of a FRP. Depending on the material combination, excellent electrical and thermal insulation properties can be achieved, which is why FRPs are used as insulators for example in housings for fuse boxes, but also as hollow insulators (Fig. 1.9). However, a high thermal or electrical conductivity can also be achieved, for example, by adding nanoparticles into the polymer matrix [22–25].

Compared to metals, FRPs also offer the advantage of radiation transparency. They can be transparent for radiation in the form of radio waves or X-rays. This can be relevant, for example, in the field of medical technology for the positioning of implants. While metallic components strongly radiate in imaging processes, such as computer tomography, and thus prevent a precise observation, FRP components sometimes even require the addition of metal markers so that they can be localized [26]. In automotive engineering new design possibilities arise, for example, when radio antennas can be installed in the interior [27, 28].

In a FRP, the polymer completely embeds the fibers and thus protects them from environmental influences. Thus, all the advantages associated with pure polymers can generally be associated with FRPs: corrosion resistance, chemical resistance and generally a high resistance to environmental influences of all kinds. Depending on the requirements, the appropriate polymer can be selected to integrate the

Fig. 1.9 Conical hollow insulators made of GFRP. Printed with permission of MR Maschinenfabrik Reinhausen GmbH

Fig. 1.10 Crash-element made of natural fiber-reinforced polymer. Printed with permission of Carl Hanser Verlag GmbH & Co. KG and Leibniz-Institut für Verbundwerkstoffe GmbH

protection against environmental influences into the material, so that no additional protective layers are required. For example, FRPs are suitable for the construction of chemical containers as well as for components in contact with corrosive media, such as salt water. In addition, biocompatible plastics for medical technology, such as implants, are available [29].

Finally, FRPs also offer the possibility of integrating the functionality of a crash element. Certain structures of fibers and matrix lead to a high energy absorption rate. They are therefore ideal crash elements. Figure 1.10 shows a crash element made of natural fiber-reinforced thermoplastic in which the forming properties of the semi-finished product were exploited to form a structure, which offers full functionality even in the event of an impact at different angles [30].

Cost reduction: The comparatively high costs for FRP components are considered to be one of the main obstacles to a broad industrial application of FRPs [9, 31]. They often result from the relatively high material costs, but partially also from manufacturing process technologies, which are immature in terms of high-volume production. However, there are also cases in which FRPs can bring cost advantages over other materials. For example, for the production of the same number of shell components, a design based on thermoset FRP molding compounds requires less tools than a metal sheet design. This can lead to cost advantages, if new tools are required on a regular basis, e.g. due to design changes (cosmetic facelifts, etc.) [27]. Exploiting these cost advantages requires an optimized selection of semi-finished products and processes for the respective application.

The above selection of specific advantages is by no means complete. For example, the minimal thermal expansion of CFRPs, the excellent acoustic damping properties of natural fiber-reinforced polymers, or the excellent surface quality of some thermoset molding compounds could be further reasons for the application of FRPs.

Apart from the advantages, there are also specific **disadvantages** associated with FRPs. For example, the material costs are often comparatively high, especially for high-performance FRPs. The same applies to development costs, whereby missing standards in terms of design methods, underdeveloped simulation capabilities and lack of material property data further complicate the development. In addition, the in-use temperature limitations for the polymer matrix often limit the fields of application. In addition, there are numerous challenges in quality assurance, repair and the recycling of FRPs [9, 32, 33].

It must be the **target of product development** with FRPs to maximize exploitation of the specific advantages of FRPs while compensating for the disadvantages. This requires the integration of different competences. It is in this context that the present book was written.

1.3   Product Development Versus Integrated Product Development (IPD)

Product creation describes the process of creating a product from the initial idea or customer order to the final delivery to the user. Development forms the core of product creation, as shown in Fig. 1.11 [34].

The development typically includes the following activities [35]:

- Conception: Search for a solution to solve the problem under consideration
- Draft: Specification of the solution by defining the design and the material
- Elaboration: Preparation of manufacturing and utilization documents.

Accordingly, the component properties are determined during the product development. It thus sets the boundary conditions for all subsequent steps (manufacturing, assembly, operation, maintenance, repair, disposal, etc.), no matter if intentional or unintentional. This often leads to problems in modern product development.

In the pre-industrial age, the production of goods was strongly influenced by the craft professions. If a blacksmith was commissioned to forge a pair of pliers, the complete development of this product was in his hands, starting from the procurement of materials and ending with manufacturing and delivery. Hence, for the blacksmith it was quite easily possible to consider the complete production process, from the beginning on. In the course of industrialization, however, products were increasingly produced in series, which made a division of tasks necessary. However, a mere division according to quantity (parallel division), where, e.g., 1000 blacksmiths each completely produce a pair of pliers is inefficient. Therefore, an additional sequential division makes sense, so that the individual workers can each concentrate on one specific subtask. This focusing brought the advantage of specialization, which made it possible, to manufacture increasingly complex products. Many of today's products, for example smartphones, are too complex to be designed or even manufactured by one person alone. Hence, division of tasks is omnipresent today. Most modern companies are divided into specialized departments, such as management, human resources, controlling, purchasing, sales, development, production planning, production, logistics and maybe even more. Even though the division of tasks is a basic prerequisite for the performance of

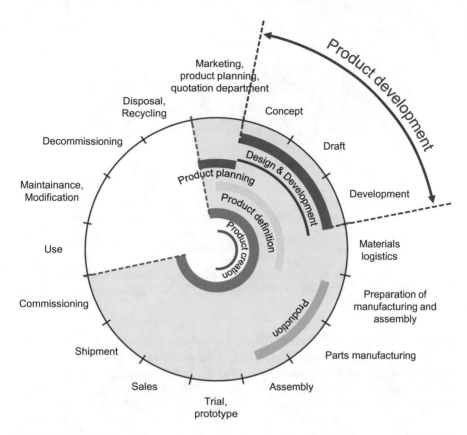

Fig. 1.11 Visualization of the product development as core of the product creation within the overall product life. Adapted from [34] (Image adapted, printed with permission of Carl Hanser Verlag GmbH & Co. KG)

today's economic systems, it is not without problems. These problems are particularly evident in product development. They are illustrated in Fig. 1.12: between the departments, "mental walls" and interface problems occur, especially regarding communication. At the same time, the individual departments are increasingly focused on optimizing themselves and not the product, i.e. they try to optimize their internal processes, in order to increase the supposed output. The resulting output is then "thrown over the wall" to the following department, which is referred to as "throw it over the wall" mentality. Problems that may arise in other departments are of secondary priority. This results in designs, which cause an enormous effort in the production development and are unnecessarily expensive to produce. The problem is further exacerbated by the fact that the strong specialization of the individual employees causes them to lose their holistic sight on the product creation. They develop their own technical language, which makes communication to other departments even more difficult and due to their limited field of activity, the

Materials logistics

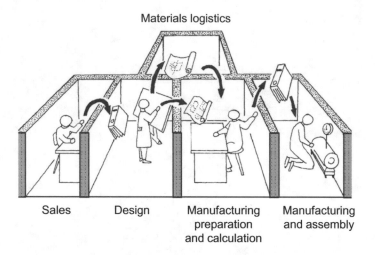

Sales   Design   Manufacturing   Manufacturing
                             preparation   and assembly
                             and calculation

Fig. 1.12 "Throw it over the wall"-mentality resulting from extensive division of tasks in conventional product development. Adapted from [34] (Printed with permission of Carl Hanser Verlag GmbH & Co. KG)

motivation is negatively affected. Hence, the holistic consideration of the product creation process is further lost. This results in retrospective design changes based on insights gained in subsequent development steps. In this context, the "rule of ten" is often referred to, which states that the costs of eliminating a design error increases by a factor of ten, with each phase of product development that has elapsed since the occurrence of the error. For example, if a design error only becomes apparent during the course of prototype production, the costs are many times higher than if the error would have already been detected during the design phase. Hence, sequential product development results in expensive and time-consuming correction loops, since not all subsequent phases of the life cycle are taken into account in the design phase. Furthermore, the general problem of not holistically (i.e. with regard to all phases of product creation and life cycle) optimized products emerges. The indispensable division of tasks therefore turns from a solution back to a problem [34, 36, 37].

Integrated Product Development (IPD) is considered to be one of the most significant and industrially most common trends in product development [38]. It aims to solve the problems resulting from the division of tasks in product development, by integrating all competences and skills, given in the specialized departments, in a joint product development (Fig. 1.13). Ehrlenspiel et al. [34] define the IPD as a target-oriented combination of organizational, methodological and technical measures used by holistically thinking product developers. This is explained in more detail in the following section.

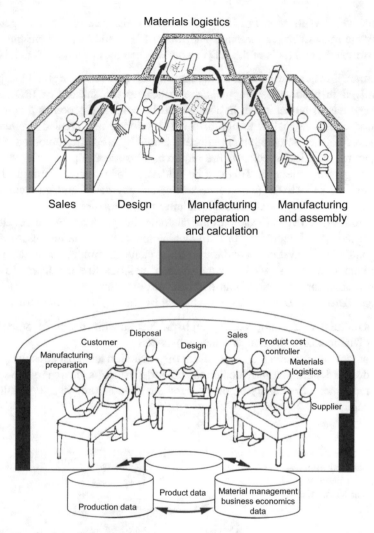

Fig. 1.13 Target of the IPD—cooperative, integrated product development. Adapted from [34] (Image adapted, printed with permission of Carl Hanser Verlag GmbH & Co. KG)

1.4   Methods of IPD

A successful IPD requires an appropriate corporate structure, which enables the implementation of the measures. Hence, the management plays a key role. Deming [39] states that 85 percent of the reasons for failure to meet customer expectations can be traced back to deficiencies in systems and processes and not so much to employees and that the role of management is therefore to change the processes not the employees. With regard to IPD, it is therefore the task of the management to identify the most effective measures and tools of IPD and to put them into action.

The measures and tools of the IPD can be assigned to four fields (Fig. 1.14) [34]:

1. **Human**: Motivation as well as common goals and knowledge of the persons involved in product development are critical to the success of IPD. A mind change has to be achieved, in a way that the company becomes a community whose common goal is to optimize the developed and created product. This requires holistic, integrative thinking and action by all parties involved. For IPD, therefore, technical specialists are required, who are equipped with the under-standing of a generalist. This can be achieved, for example, through the inte-gration of objectives (e.g. success-oriented payment) and the creation of inclusive knowledge (e.g. through training or job rotation).
2. **Methodology**: Typical examples are customer integration into the development process or task integration for individual employees (extending responsibilities beyond classic system boundaries). This includes simple methods, such as working in teams, as well as sophisticated methods like the Japanese Kaizen which aims for continuous improvement of processes [40].
3. **Organization**: IPD must also be reflected by the organizational structure by

    a. structural integration, e.g. through product-specific forms of organization, divisional organization, flat hierarchies etc.
    b. process integration, e.g. in the form of production and cost consulting for the design department provided by the other departments, design reviews, failure mode and effect analysis, or the parallelization of processes ("simultaneous engineering").

Fig. 1.14 Four components of IPD. Adapted from [34] (Printed with permission of Carl Hanser Verlag GmbH & Co. KG)

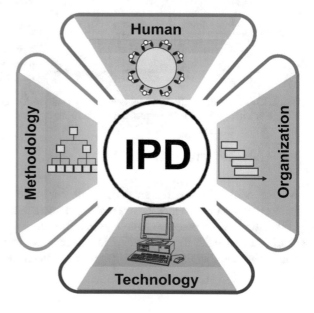

   c. spatial integration through the establishment of development centers and
      joint workspaces for the specialists from different departments.

4. **Technology**: The employees involved in the development process must have
   access to the same data and use the same tools wherever possible. In order to be
   able to take the entire product life cycle into account, already in the design
   phase, technologies for the early recognition of properties, such as simulations
   or rapid prototyping, are also necessary.

There are numerous approaches to the exact application of the measures, and
many more examples could be added to those mentioned above. Hence, every
person involved in a product development process at some point most likely will be
part of a measure attributable to IPD. However, success depends on whether these
measures are actually put into action properly. It is of crucial importance that these
measures are implemented by holistic thinking product developers. This book
intends to contribute to a holistic perspective. For this, it is helpful to explain why
the IPD is of particular relevance for FRPs, which is presented in the next section.

1.5   Relevance of IPD for FRPs

Figure 1.15 shows two vehicles that, at first glance, do not seem to have much in
common: On the left, the Trabant,[2] which was manufactured in East Germany from
1950 onwards. On the right the i3[3] produced by BMW since 2013.

The common feature of both vehicles is the industrial use of FRPs. Due to a lack
of raw materials in the German Democratic Republic, the Trabant's outer skin was
made of phenolic resin impregnated cotton and was mounted on a metallic base
structure [41, 42]. The BMW i3, on the other hand, provides a CFRP passenger cell
mounted on an aluminum chassis (Fig. 1.16). The outer skin mainly consists of
non-reinforced thermoplastic components.

Even though the boundary conditions and the motivation for the use of FRPs
were different, in both cases pre-existing knowledge was limited. Thus, the
developers of both cars had to work intensively on topics related to the series
manufacturing of FRPs:

- Which **materials** are suitable?
- Where can the required materials be **procured** in sufficient quantities?
- How must the components be **designed** to withstand the loads?
- How can FRP components be **joined** to other components?
- Which **processing** methods are cost-efficient?
- What **process speeds** can be achieved?
- How to implement **quality assurance**?

---

[2]The depicted model P601 was built from 1964 to 1990.
[3]The depicted model is a BMW i3 120Ah from 2019.

Fig. 1.15 Trabant 601 and BMW i3—two well-known examples for the industrial application of FRPs. Left image: image adapted from „Trabant 601 S" from, Flominator, License for use: https://creativecommons.org/licenses/by-sa/3.0/deed.de; right image: Copyright of BMW AG

Fig. 1.16 Trabant with FRP panels on a metallic structure and BMW i3 with CFRP passenger compartment on aluminum chassis. Right image: Copyright of BMW AG

- How **resistant** are the components in use to environmental influences?
- How can components be **repaired** or replaced in case of damage?
- How can **sustainable** production be achieved?
- How to deal with **end-of-life components** and **production waste**?

Hence, questions arise along the entire product life cycle. Of course, these questions also arise when other materials, such as metals, are used. However, for FRPs, they provide very particular challenges, which require a holistic, i.e. integrated product development. The VDI—guideline 2014 on the "development of components made of FRP composites" [43] states:

"Interdisciplinary work between the disciplines of design, materials, manufacturing, calculation and economy is essential".

Four points are given as reasons:

1. In principle, in addition to a **geometrical** design, a design to **material** and **manufacture** is required.

2. The **availability** of **reliable material parameters** for dimensioning is problematic.
3. **Manufacturing restrictions** determine the design freedom of the designer; Neglecting this causes higher production effort and higher costs.
4. The **production technology** determines the **component quality** and **reproducibility**.

The last point deserves special emphasis as it is a specific feature of FRPs: The FRP, and thus its final properties as a material, only emerge after the actual component production. This might apply to other materials as well, but the heterogeneous structure adds particular complexity and thus variation to the manufacturing processes. Another important factor is that FRPs form a comparatively young material group, which is characterized by a great diversity in many areas and an equally rapid development, causing the diversity to quickly increase. The questions raised during the development of the Trabant and the BMW i3 were the same, but the range of possible answers was highly different. New fiber materials, matrix polymers and semi-finished products are constantly pushing their way into the already confusing market. Cost differences of several 1000% are quite common. The diversity among the manufacturing processes is also constantly increasing, whereas different goals, such as high quality or productivity, can be in focus. The cost efficiency of the processes therefore strongly depends on the boundary conditions and what is more, the processes have a particularly strong influence on the mechanical performance of the produced FRPs and the corresponding design limitations. Since the actual component properties are thus a result of the specific combination of materials and processes, material property values from data sheets can only be used for component design to a limited extent. In addition, even for the design methods, the failure models and the characterization methods for material properties a stagnation of research and development activities is not in sight. The constant change continues to extend over the entire FRP process chain up to the topics of repair and recycling, which are currently receiving a lot of attention.

In short: Product development with FRPs requires extensive knowledge in different areas, but especially with regard to interdependencies. An impossible task for a single individual. Hence, due to the strong interdependencies, the risks associated with division of tasks in product development are even more remarkable compared to other material groups. The guiding principles of IPD are therefore of particular relevance:

- **Team-oriented** and **interdisciplinary work** within the company and with customers and suppliers
- Consideration of the entire **product life cycle**
- Attention to the **interdependencies** between the **elements of the product life cycle**

The consistent implementation of an IPD also gives the opportunity to exploit the enormous potential, offered by FRPs, in the best possible way. This book intends to help the reader with basic knowledge about product development with

FRPs and interdependencies along the product life cycle. Furthermore, it is shown how an IPD can be practically implemented.

1.6   Questions for Self-check

In the following, you will find questions and tasks that will help you to reflect on the most important contents of this section. The solutions can be found in Chap. 7.

R1. Give five reasons for using FRPs.
R2. Give two reasons for division of tasks.
R3. Name two possibilities for division of tasks.
R4. Define "integrated product development."
R5. Name the four elements of integrated product development according to Ehrlenspiel.
R6. Name four reasons why integrated product development is especially relevant for FRPs.

Literature

1. Griffith, A.A., Eng, M.: The phenomena of rupture and flow in solids. Philos. Trans. R. Soc. Lond. A **221**(582–593), 163–198 (1921)
2. Gordon, J.E.: The new science of strong materials: or why you don't fall through the floor. Penguin UK, London (1991)
3. Schürmann, H.: Konstruieren mit Faser-Kunststoff-Verbunden. Springer, Berlin (2007)
4. Neitzel, M., Mitschang, P., Breuer, U.: Handbuch Verbundwerkstoffe: Werkstoffe, Verarbeitung Anwendung. Carl Hanser Verlag GmbH Co KG, Munich (2014)
5. Gao, S.-L., Mäder, E.: Characterisation of interphase nanoscale property variations in glass fiber reinforced polypropylene and epoxy resin composites. Compos. A Appl. Sci. Manuf. **33** (4), 559–576 (2002)
6. Puck, A.: Festigkeitsanalyse von Faser-Matrix-Laminaten: Modelle für die Praxis. Hanser, Munich (1996)
7. Witten, E., Mathes, V., Sauer, M., Kühnel, M.: Composites-Marktbericht 2018: Marktentwicklungen, Trends, Ausblicke und Herausforderungen. AVK—Industrievereinigung verstärkte Kunststoffe e. V./Carbon Composites e. V. (2018)
8. Witten, E., Sauer, M., Kühnel, M.: Composites-Marktbericht 2017: Marktentwicklungen, Trends, Ausblicke und Herausforderungen. AVK Industrievereinigung verstärkte Kunststoffe e. V./Carbon Composites e. V. (2017)
9. Lässig, R., Eisenhut, M., Mathias, A., Schulte, R.T., Peters, F., Kühmann, T., Waldmann, T., Begemann, W.: Series production of high-strength composites—Perspectives for the german engineering industry. Roland Berger Strategy Consultants (Munich) (2012)

10. Canyon GmbH: Vertical Comfort Lateral Stiffness. Downloaded from https://www.canyon.com/en/innovation/vcls/, downloaded on 29.07.2018 (2018)
11. Schimmer, F., Motsch, N., Hausmann, J., Magin, M., Bücker, M.: Analysis on formed bolted joints for thick-walled CFRP in wind power industry. In: 21st International Conference on Composite Materials, Xi'an, China, 20–25.08.2017 (2017)
12. Windkraft Journal: Torsionswelle Flexshaft 5.0 gewinnt Innovationspreise in Paris und Hannover (Article from 07.05.2013). Downloaded from https://www.windkraft-journal.de/2013/05/07/torsionswelle-flexshaft-5-0-gewinnt-innovationspreise-in-paris-und-hannover/39686, downloaded on 29.07.2018 (2013)
13. Breuer, U.P.: Commercial Aircraft Composite Technology. Springer, Berlin (2016)
14. Wiedemann, M., Sinapius, M.: Adaptive, Tolerant and Efficient Composite Structures. Springer Science & Business Media, Berlin (2012)
15. Friedrich, K., Breuer, U.P.: Multifunctionality of Polymer Composites: Challenges and New Solutions. William Andrew (Elsevier), Oxford and Waltham (2015)
16. Klingler, A., Sorochynska, L., Wetzel, B.: Toughening of glass fiber reinforced unsaturated polyester composites by core-shell particles. Key Eng. Mater. (2017)
17. Chapple, S., Anandjiwala, R.: Flammability of natural fiber reinforced composites and strategies for fire retardancy: a review. J. Thermoplast. Compos. Mater. **23**(6), 871–893 (2010)
18. Hannemann, B., Backe, S., Schmeer, S., Balle, F., Breuer, U., Schuster, J.: Hybridisation of CFRP by the use of continuous metal fibers (MCFRP) for damage tolerant and electrically conductive lightweight structures. Compos. Struct. **172**, 374–382 (2017)
19. Wetzel, B., Haupert, F., Zhang, M.Q.: Epoxy nanocomposites with high mechanical and tribological performance. Compos. Sci. Technol. **63**(14), 2055–2067 (2003)
20. Semar, J., May, D., Mitschang, P.: Evaluation of different perforation patterns for laminate-integrated heating foils in wind turbine rotor blades. In: 18th European Conference on Composite Materials, Athens, Greece (2018)
21. Gurka, M.: The Physics of Multifunctional Materials: Concepts, Materials, Applications. DEStech Publications Inc., Lancaster (2019)
22. Pleşa, I., Noţingher, P.V., Schlögl, S., Sumereder, C., Muhr, M.: Properties of polymer composites used in high-voltage applications. Polymers **8**(5), 173 (2016)
23. Al-Oqla, F.M., Sapuan, S.: Natural fiber reinforced polymer composites in industrial applications: feasibility of date palm fibers for sustainable automotive industry. J. Clean. Prod. **66**, 347–354 (2014)
24. Hildebrandt, K., Mitschang, P.: Effect of incorporating nanoparticles in thermoplastic fiber reinforced composites on the electrical conductivity. In: 18th International Conference on Composite Materials, Jeju Island, South Korea, 21–26.08.2011 (2011)
25. Chen, H., Ginzburg, V.V., Yang, J., Yang, Y., Liu, W., Huang, Y., Du, L., Chen, B.: Thermal conductivity of polymer-based composites: fundamentals and applications. Prog. Polym. Sci. **59**, 41–85 (2016)
26. Plastics Today: Plastic clip advances treatment of aneurysms (Article from 22.11.2016). Downloaded from https://www.plasticstoday.com/medical/plastic-clip-advances-treatment-aneurysms/176833759246139?cid=flyout, downloaded on 09.09.2018 (2016)
27. Sommer, M.: Chancen für SMC und BMC im Automobilbau. 10. Internationale AVK-Tagung für verstärkte Kunststoffe und technische Duroplaste, Stuttgart: 5–6. November 2007 (2007)
28. Ernstberger, U., Weissinger, J., Frank, J.: Mercedes-Benz SL: Entwicklung und Technik. Springer Fachmedien Wiesbaden (2013)
29. Asghari, F., Samiei, M., Adibkia, K., Akbarzadeh, A., Davaran, S.: Biodegradable and biocompatible polymers for tissue engineering application: a review. Artifi. Cells, Nanomed. Biotechnol. **45**(2), 185–192 (2017)
30. Kunststoffe International: Thermoplastischer offaxisstabiler Crashabsorber—Crash-Muffin aus Faser-Kunststoff-Verbunden (Article from 03.11.2014). Downloaded from https://www.kunststoffe.de/produkte/uebersicht/beitrag/thermoplastische-crash-muffins-crashabsorber-aus-faser-kunststoff-verbunden-FRP-945632.html, downloaded on 09.09.2018 (2014)

31. Eickenbusch, H., Krauss, O.: Kohlenstofffaserverstärkte Kunststoffe im Fahrzeugbau—
    Ressourceneffizienz und Technologie. VDI Zentrum Ressourceneffizienz GmbH, Berlin
    (2013)
32. Oliveux, G., Dandy, L.O., Leeke, G.A.: Current status of recycling of fiber reinforced
    polymers: review of technologies, reuse and resulting properties. Prog. Mater Sci. **72**, 61–99
    (2015)
33. Königsreuther, P.: MaschinenMarkt: Endlosfaserverstärkte Thermoplaste haben ein öffentliches
    Forum (Article from 14.07.2015). Downloaded from https://www.maschinenmarkt.vogel.de/
    endlosfaserverstaerkte-thermoplaste-haben-ein-oeffentliches-forum-a-497623/, downloaded on
    31.05.2019 (2015)
34. Ehrlenspiel, K., Meerkamm, H.: Integrierte Produktentwicklung: Denkabläufe,
    Methodeneinsatz, Zusammenarbeit. Carl Hanser Verlag GmbH Co KG, Munich (2013)
35. Pahl, G., Beitz, W., Schulz, H.-J., Jarecki, U.: Pahl/Beitz Konstruktionslehre: Grundlagen
    erfolgreicher Produktentwicklung, Methoden und Anwendung. Springer-Verlag, Berlin/
    Heidelberg (2013)
36. Schmitt, R., Pfeifer, T.: Qualitätsmanagement: Strategien–Methoden–Techniken. Carl Hanser
    Verlag GmbH Co KG, Munich (2015)
37. Komorek, C.: Integrierte Produktentwicklung: der Entwicklungsprozeß in mittelständischen
    Unternehmen der metallverarbeitenden Serienfertigung. Erich Schmidt Verlag GmbH &
    Co KG, Berlin (1998)
38. Gerwin, D., Barrowman, N.J.: An evaluation of research on integrated product development.
    Manage. Sci. **48**(7), 938–953 (2002)
39. Deming, W.E.: Out of the Crisis: Quality, Productivity and Competitive Position. Cambridge
    University Press, Cambridge (1986)
40. Kaizen, I.M.: The Key to Japan's Competitive Success. MacGraw-Hill, New York (1986)
41. Zepf, H.P.: Faserverbundwerkstoffe mit thermoplastischer Matrix. Expert-Verlag, Renningen/
    Malmsheim (1997)
42. Faruk, O., Tjong, J., Sain, M.: Lightweight and Sustainable Materials for Automotive
    Applications. CRC Press, Boca Raton (2017)
43. Verein Deutscher Ingenieure e. V.: VDI-Richtlinie 2014: Entwicklung von Bauteilen aus
    Faser-Kunststoff-Verbund (Blatt 1: Grundlagen, 1998, Blatt 2: Konzeption und Gestaltung,
    1993, Blatt 3 Berechnungen, 2006) (2006)

Chapter 2
Implementation of IPD

Abstract This chapter shows how a development team could be structured in order to perform integrated product development (IPD). Subsequently, a procedure for the IPD of a FRP component is described, including a corresponding division of tasks within the team.

Keywords Integrated product development · Team structure · Division of tasks

2.1 Structure of the Development Team

The integrated development of a FRP component requires the cooperation of different experts, bringing together their specialized knowledge, in order to be able to provide the necessary expertise for a holistic and optimal solution. Accordingly, the establishment of a team is indispensable, which raises the question of the ideal combination of team members. First, it should be noted that expertise in the fields of

- design,
- manufacturing,
- material science/technology and
- business economics

must be integrated into the team. Of course, there are engineers who provide special competences in more than one field. However, with regard to the high degree of diversification in the field of FRPs, as discussed in the last section, it can be assumed that commonly a development team should consist of more persons. First of all, a design engineer carrying out the actual component design. Furthermore, a manufacturing engineer, selecting the corresponding manufacturing process and consulting with the design engineer concerning the design requirements resulting from this selection. Due to the strong interdependencies between process and materials, manufacturing engineers often also provide profound expertise in terms of standard materials and semi-finished products. Yet, due to the high variety of materials, special requirements to the material justify the additional integration of a

D. May, *Integrated Product Development with Fiber-Reinforced Polymers*, Engineering Materials, https://doi.org/10.1007/978-3-030-73407-7_2

material specialist. In order for the manufacturing engineer to be able to select and design the optimum process, he must have expertise in the field of business economics. However, detailed cost analysis, e.g. for make-or-buy decisions, or even a holistic life cycle analysis including economic, technical, social and ecological aspects, require a higher level of specialization, which is why the integration of a team member with special expertise in business economics is often also necessary. Regardless of the team size, clear structures are necessary. Even with small teams, a team leader must be appointed. Since the tasks of the design engineer are at the center of the development, for smaller teams it can make sense to appoint him as project manager, however, the availability of leadership skills is of particular importance. Table 2.1 shows in simplified form the tasks and their distribution in a minimal team.

Table 2.1 Possible division of tasks in small-sized IPD team

Team member 1 Department: Design	Team member 2 Department: Manufacturing	Team member 3 Department: Materials
• Team leader • Main contact for customer • Planning and moderation of the coordination meetings • Identification of the design requirements • Merging of all requirements of all departments in an overall requirements catalog • Conception, drafting and pre-dimensioning • Selection of fiber material and structure of fiber reinforcement • Revision of the concept considering the results of the other team members • Organization of the tests for the determination of the material properties • FE-design and—optimization • Component test • Technical evaluation and comparison with requirements catalog • Techno-economic and strategic evaluation • Definition of recommended actions for customer • Preparation of production documents • Strength verification	• Participation in coordination meetings • Identification of the process requirements • Selection of a suitable manufacturing process • Elaboration of a process concept • Selection of specific types of semi-finished products • Cycle time estimation • Management of process simulation • Economic process analysis • Management of prototyping • Comparison of manufacturing concept with requirements	• Participation in coordination meetings • Identification of the material requirements • Check of material selections of other departments • Decision for thermoplastic or thermoset matrix (together with manufacturing department) • Selection of specific matrix polymer and possible modification needs • Comparison of material concept with requirements

The large number of tasks in the development process allows a further division and the integration of additional team members. This offers the advantage of combining more competencies and achieving a deeper specialization. In addition, increased parallelization can reduce the development time. The target-oriented size of the team thus depends above all on the complexity of the component to be developed and the time constraints. However, effective cooperation in the team must be ensured, which becomes more complicated as the number of team members increases. Therefore, for larger teams or time-intensive projects a full-time project manager, with appropriate management skills, is appropriate. Table 2.2 presents a possible distribution of tasks in an extended project team. Such a strong division of tasks is useful in case of a high project complexity, or if the team members are working in several projects at the same time and thus an adequate workload is ensured.

2.2   Procedure and Division of Tasks for IPD with FRPs

In the following section, a fundamental and idealized approach for the IPD of a FRP component is proposed. The overall structure of this book is based on this procedure. It is assumed that the team is divided into representatives of the design, manufacturing and materials departments. Although a further subdivision will very often be useful, this subdivision is ideal for presenting the approach because it offers a good balance between the requirement to present interdisciplinary cooperation and the effort to ensure the best possible clarity. In addition, this approach comprises of a higher-level decision-maker who corresponds to the customer.

Figures 2.1 and 2.2 schematically show the basic design of the IPD as proposed in this book.

In the following section, this fundamental concept is briefly explained in order to give the reader an overview. In the following sections of the book, the single phases of the IPD are then presented in detail and the knowledge necessary for implementation is provided.

Phase 1: Preparation of requirements catalog

Step 1-1: Kick-off meeting with the customer

The initial meeting with the customer marks the beginning of every product development project. Thereby, the term customer can be defined in different ways (e.g. internal or external). Regardless of the exact definition, it is necessary to collect all requirements for the FRP product to be developed. Not only the end properties of the product but also the boundary conditions for the complete product life cycle have to be identified. This first step is therefore critical, which is why participation of all departments involved in the development, as well as the higher-level decision-maker is mandatory. The higher-level decision-maker can be a representative of the customer or a person within the company entrusted with this task.

Table 2.2 Possible division of tasks in an extended IPD team

Team leader			
- Main contact for customer - Project management - Planning and moderation of coordination meetings - Merging of results in an overall techno-economic evaluation - Definition of recommended actions for customer			
Sub-team Design	**Sub-team Manufacturing**	**Sub-team Materials**	*Technical purchasing*
Sub-team leader	*Sub-team leader*	*Sub-team leader*	
• Management of the sub-team • Representation of design department in coordination meetings • Identification of the design requirements • Comparison of design concept with requirements	• Management of the sub-team • Representation of manufacturing department in coordination meetings • Identification of the process requirements • Comparison of process concept with requirements	• Management of the sub-team • Representation of materials department in coordination meetings • Identification of the material requirements • Comparison of selected materials with requirements	• Support of project team • Verification of market availability and prices for semi-finished products
Design engineer	*Process engineer*	*Materials engineer*	
• Conception, drafting and pre-dimensioning • Selection of fiber material and structure of fiber reinforcement • FE-design and –optimization • Preparation of production documents	• Selection of a suitable manufacturing process • Elaboration of a manufacturing concept • Selection of specific types of semi-finished products	• Decision of thermoplastic or thermoset matrix (together with manufacturing department) • Selection of specific matrix polymer and possible modification needs	
Calculation engineer	*Economist*	*Test engineer*	
• Strength verification	• Economic process evaluation	• Determination of material properties (coupon-level) • Component testing	
	Simulation expert		
	• Process simulation for cycle time estimation		
	Prototyping		
	• Manufacturing of prototypes and material samples		

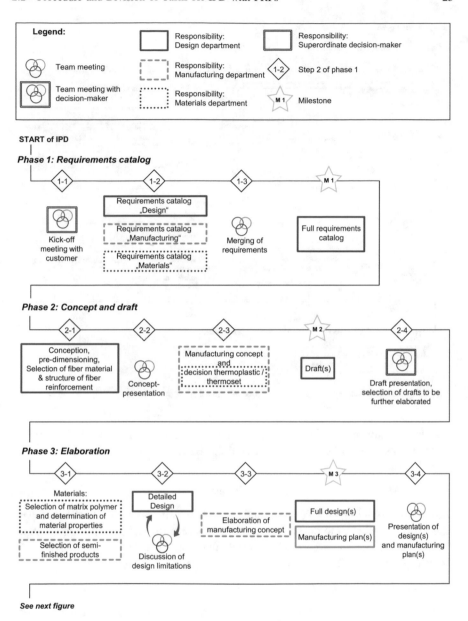

Fig. 2.1 Idealized concept for integrated product development—part 1

Step 1-2: Requirements catalogs of the departments

Considering the team structure, requirements catalogs for the areas "design," "manufacturing" and "materials" are defined, which is comprised of the subareas relevant to them and may be partially redundant. Each department interprets the

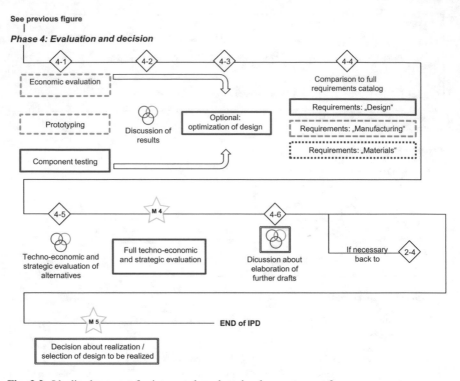

Fig. 2.2 Idealized concept for integrated product development—part 2

customer's information with its own specific background and "translates" them into requirements relevant to their respective field. Hence, requirements might not only be redundant, but also contradictory.

Step 1-3: Merging the requirements catalogs

The various requirements catalogs must be merged, involving each of the departments. In particular, the design department must ensure that indirect requirements for the design are identified, which result from the requirements catalogs of the other departments. Any contradictions must be resolved.

Milestone 1: Overall requirements catalog

The end of the first phase is the overall requirements catalog, which is defined taking into account the results of the previous coordination meeting. The final elaboration of the catalog should be assigned to the design department.

Phase 2: Concept/Draft phase

The IPD intends to achieve improved coordination through extensive parallelization of tasks. Thereby, an efficient working method requires that a profound working basis for the various departments is created relatively quickly. Thus, the concept development is initially carried out based on rough calculations, so that, for example, the fundamentals for the manufacturing concept development are available at an early stage.

Step 2-1: Concept development, pre-dimensioning, selection of fiber material and definition of the structure of the fiber reinforcement

The second phase begins with the definition of the critical load cases, from which the requirements for materials and geometry can be derived. Solution concepts are then developed taking into account the requirements defined in phase 1. Each developed concept includes a specific fiber material combined with a suitable reinforcement structure as well as a first, pre-dimensioned design draft that can be used for the development of the manufacturing concept. The concept is developed under the responsibility of the design department team member.

Step 2-2: Concept presentation

The concept forms the basis for the work in the different departments and is therefore presented and discussed with the other team members in a coordination meeting.

Step 2-3: Process concept and decision thermoset/thermoplastic

The requirements catalog and the component design serve the team member from the manufacturing department as a basis for the selection of the manufacturing process. The decision whether to choose a thermoset or thermoplastic matrix polymer is also made at this stage without specifying the specific matrix polymer. The decoupling of the decision for the polymer category from the decision for a specific polymer is because the decision between thermoset and thermoplastic has far-reaching consequences for the development of the design and the manufacturing concept. This decision should therefore be made as early as possible in order to enable simultaneous work in the different departments.

Milestone 2: Draft (drafts)

At this point, one or more alternative drafts are available, each being defined by the selected fiber material, the structure of the fiber reinforcement, the selected polymer class (thermoplastic or thermoset), an initial, pre-dimensioned design draft and a selected manufacturing process (group). The individual drafts can be fundamentally different or only differ in a single point, e.g. the manufacturing process.

Step 2-4: Presentation of the design, determination of the designs to be worked out

In a further coordination meeting, the drafts are compared with each other and, involving the customer, it is decided which drafts should be further elaborated. Since the following steps go together with enormous effort, it is usually not reasonable to follow up on all drafts, since in the end only one of the drafts is actually implemented. On the other hand, reducing the number of drafts to be elaborated also increases the danger that eventually none of the selected drafts will lead to the desired result. In this case, one of the drafts that were not prioritized in the first place would have to be elaborated, resulting in a loss of time. The risk of this loss of time and the resources required for each draft to be prepared must therefore be carefully balanced.

Phase 3: Elaboration

The elaboration of the chosen draft(s) starts within the departments.

Step 3-1: Elaboration of the material concept

Based on the selected manufacturing process and the other boundary conditions the team member from the manufacturing department selects suitable semi-finished products. The task of the team member from the materials department is to identify a suitable matrix polymer, which meets the requirements defined in the requirements catalog and is available as the selected type of semi-finished product. Furthermore, it must be suitable for the selected manufacturing process in terms of processing behavior, which, if necessary, must be investigated using appropriate test methods. After all decisions have been made on the material side, material characteristics are determined as a basis for the design.

Step 3-2: Elaboration of the design and coordination of the design restrictions

The detailed design is a decisive step of the product development. At this point, the requirements from all areas and from all phases of the product life cycle must be taken into account. This is the only way to achieve a holistically optimized design that is, among other things, suitable for manufacturing, for joining, for repair and resource-efficient. The final shape of the component is determined during this step.

Step 3-3: Elaboration of the manufacturing concept

After all materials and the final shape of the component have been determined, the elaboration of the manufacturing concept can take place. For this purpose, the exact process variant is selected and then the corresponding necessary equipment is identified and selected. This is followed by the process design, in which the system configuration is optimized based on a cycle time estimation. Furthermore, a more precise cycle time prediction can be made by a process simulation and the tool and process parameters can be optimized on this basis. If necessary, optimization measures for the component design and semi-finished products are derived from the results. Finally, suitable methods for quality assurance are identified.

Milestone 3: Construction and production plans

The third milestone is reached when the technical drawings for the component, the specifications for the semi-finished products and the manufacturing plans including specifications for the production systems, etc. are available.

Step 3-4: Presentation of the design(s) and production plans

At the end of the elaboration, the designs and manufacturing plans are discussed within the team, to clarify final questions and to assess benefits and costs of possible design modifications.

Phase 4: Evaluation and decision

If several drafts have been prepared, or drafts based on other materials are available (e.g. in the context of metal substitution), a decision must be made, which of the alternatives should be implemented. This requires an assessment as a basis for the decision-making.

Step 4-1: Economic process analysis, prototyping, and component testing

Economic efficiency is one of the most important factors in decision-making. For this reason, process-based cost modeling is first carried out, to calculate the manufacturing costs resulting from the alternatives. Due to the necessary process expertise, ideally the team member from the manufacturing department is responsible for this task. In order to validate the requirements, prototypes are manufactured (responsible: manufacturing engineer) and tested (responsible: design engineer).

Step 4-2: Discussion of results

The results of the cost analysis and the component tests are discussed by the team and the costs and benefits of possible design adjustments are evaluated.

Step 4-3: Optimization of the design

If the benefit of a design modification is considered to be greater than its costs (including time consumption), the implementation is carried out by the design department.

Step 4-4: Comparison with requirements catalogs for "design," "production" and "materials"

A comparison of each concept with the requirements catalog can take place after the prototypes were tested, the component properties were derived from the simulation and prototype/material tests, and after the boundary conditions and costs of manufacturing were derived from the process simulation and cost analysis.

Step 4-5: Techno-economic and strategic evaluation of the alternatives

In a joint team meeting, the achievement of development targets is systematically evaluated in terms of technical, economic and strategic target criteria. By weighing up the targets, a quantitative comparison of the achievement of targets is obtained. The joint elaboration of the evaluation in a workshop is useful, to integrate the different perspectives on the achievement of targets and thus avoid a subjective evaluation.

Milestone 4: Techno-economic and strategic overall evaluation

The achievement of a complete evaluation of all alternatives marks the fourth milestone.

Step 4-6: Voting: Implementation or elaboration of further drafts?

Together with the customer, a decision must be made at the end, whether a further draft is to be prepared, e.g. because an overall better achievement of objectives is expected with a changed implementation risk. If further drafts are to be developed, it must be decided which drafts are to be developed. The procedure is therefore repeated from step 2-4.

When the overall evaluations for all alternatives are available and no further drafts are developed, then a decision must be made together with the customer, whether one of the alternatives should be implemented and if so, which one.

Milestone 5: Decision on implementation or selection of the alternatives to be implemented

The final milestone is the decision on whether an implementation should take place and, if yes, which alternative forms the basis for it.

It will not always be possible or practical to strictly adhere to this idealized approach, in terms of both timing and the structure of the development team. However, the approach gives an example how integration can be achieved in the context of product development. The following sections therefore follow this procedure and show step by step how the individual tasks can be mastered. The knowledge required for this purpose will be provided along the way.

Chapter 3
Phase 1: Defining the Requirements Catalog

Abstract Before the actual development one has to face the challenge to comprehensively define the development task. This means that all requirements for the later product and its complete product life cycle must be identified, quantified or qualified and eventually collected into a requirements catalog. This chapter initially describes different types of requirements. Then, possible sources for the requirements are identified and the most important risks during requirements definition are described. Furthermore, helpful tools for the preparation of the requirements catalog are presented. Finally, a requirements catalog, especially for FRP products is introduced and illustrated by means of an application example.

Keywords Requirements catalog · Scenario Technique · List of main characteristics · Functional structures

3.1 Overview

Figure 3.1 shows the procedure for the creation of the requirements catalog (phase 1).

3.2 Types of Requirements and Their Sources

DIN 69905 defines the requirements catalog as a "list of requirements, which are to be fulfilled in order to achieve a desired project goal." In this context, the term requirement is defined by the quality management standard DIN EN ISO 9000:2015 as "a requirement or expectation that is specified, usually presupposed or obligatory." A specified requirement is one that is given in a "documented" form (e.g. in a requirements catalog). "Usually" in this context means that the prerequisite of a requirement is common practice. As shown in Fig. 3.2, requirements can be divided into demands and wishes.

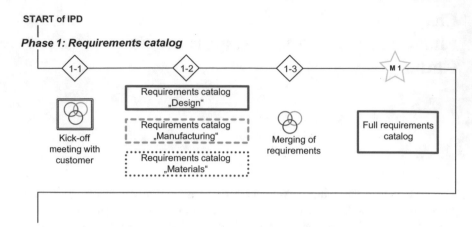

Fig. 3.1 Overview of phase 1—defining the requirements catalog

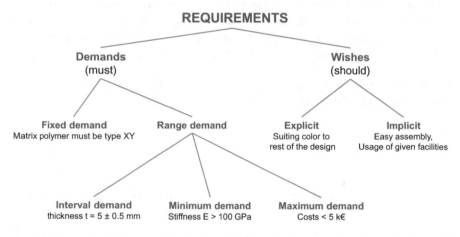

Fig. 3.2 Types of requirements. Adapted from [1] (extended by topic-related examples) after [2] (Image adapted, printed with permission of Carl Hanser Verlag GmbH & Co. KG.)

Demands are requirements that must be fulfilled by the product. There are different types of demands, e.g. fixed demands. For example, for FRP components in aircraft construction, this includes the specification of a certain aviation-certified resin system as a matrix polymer. Alternatively, range demands can be specified which define a permissible value interval (interval demand, e.g. component thickness between 4.5 and 5.5 mm), a minimum value (minimum demand: stiffness greater than 100 GPa) or a maximum value (maximum demand: manufacturing costs below 5000 €) [1, 3].

Wishes are requirements that should be fulfilled if possible. If necessary, additional expenditure might also be acceptable here. Explicit wishes are considered to be wishes that are explicitly specified by the customer, just like the demands. Content and form can correspond to a demand. The only difference is whether fulfillment is mandatory or optional. It can be useful to add a priority level to wishes (e.g. important and less important), for the case that all requirements are fulfilled and there is potential to fulfill some but not all wishes. This can be illustrated by a simple example: Suppose you want to design a seat shell, with a given weight maximum of 5.5 kg. After fulfillment of all requirements, the weight is 4.5 kg. Several wishes are still pending, including an integrated seat heating (+0.8 kg) and an adaptive lumbar support (+0.9 kg). Although the wishes are clear, there is a decision to be made by the design team, for which it has no decision basis when there is a lack of prioritization [1, 3].

In addition to the explicit wishes there are also so-called implicit wishes [1]. These are wishes that are not explicitly stated by the customer, for example, because they are obvious to the customer, which must not necessarily apply to the design team. For example, the desire for the simplest possible assembly may be a relatively obvious desire. However, if a customer has no experience with FRP, he might assume that assembly with additional components can be simply done by screwing without any problems. He therefore might not express this as a wish, although it may be critical for the FRP design. Especially during product development with FRPs, the design team must therefore recognize implicit wishes and transfer them into explicit wishes or even demands.

In order to specify requirements, the legitimate sources must first be identified. As shown in Fig. 3.3, the assumption of a simple customer-contractor situation, in which the customer himself is the source of all requirements, gives an incomplete image. Requirements can arise from several sources. Furthermore, the term customer must be understood in an abstract way. It can be a specific customer who has commissioned the product development. However, in this context a specific customer can also be given by a clearly defined market segment for which standardized requirements exist (e.g. automotive compact car segment). Alternatively, the customer can also be of anonymous nature, in the form of in-house sales defining a task without having a specific customer. Furthermore, this also includes the results of product management, derived from a defined market segmentation [4].

3.3 Risks in the Preparation of the Requirements Catalog

Collecting requirements may seem trivial at first glance. However, this step involves various risks, which are highly crucial given the critical relevance of the requirements catalog for the development process. The members of the development team should therefore constantly be aware of these risks when defining the requirements catalog. When requirements are defined, there is always the danger that one of the following applies to them:

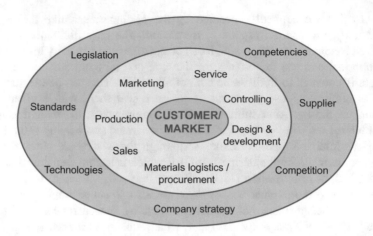

Fig. 3.3 Sources of requirements. Adapted from [1] after [5] (Image adapted, printed with permission of Carl Hanser Verlag GmbH & Co. KG)

- **Incomplete**: Probably the greatest risk regarding a requirement is its absence in the requirements catalog. For example, missing requirements with regard to media resistance can lead to an unsuitable material selection eventually limiting the components functionality and usability. In addition to the complete absence of a requirement, there is also the risk that an individual requirement is incomplete. This would be the case, for example, if maxima and minima of the service temperature range were given, but without defining which temperatures must be withstood permanently and which as short-term peaks. Here too, there is a risk of unsuitable material selection.
- **Inaccurate**: The accuracy of a requirement has a direct effect on the design freedom. Inaccurate requirements limit the number of possible solutions, while accuracy has the opposite effect. It is important that the design freedom is not unnecessarily restricted. Requirements should therefore be as accurate as possible. For example, instead of merely specifying a minimization of the installation space as a requirement, the addition of a minimum value below which a further reduction in installation space has no further technical advantages can be a valuable addition.
- **Incorrect**: A requirement defined by the customer can simply be wrong for many reasons, such as lack of expertise or care, e.g. defining false corrosive influences. However, requirements are often also untrue because the customer "overshoots the mark" and, for example, demands a higher operating temperature to be withstood than is actually necessary for the later application. Requirements should therefore always be critically reviewed.
- **Inadequate**: Especially in the field of FRP components, product developers often see themselves confronted with inadequate requirements, especially when it comes to substitution of an existing metal component. If, for example, the cost targets of a sheet metal design are set for a transport box, the FRP design is often

without a chance, although it might compensate for the higher costs by an additional benefit (more payload for the same total weight). Hence, the requirements defined by the customer should be accordingly reviewed and, if necessary, jointly revised.

- **Impossible**: High expectations are often associated with the use of FRPs, especially with regard to the potential for lightweight design. At the same time, especially in the industrial sector, cost-effectiveness is usually a top priority. For this reason, customers often define requirements that can be classified as impossible. It is important to dampen the expectations to a realistic level. The early identification of such "show stoppers," which are characterized by very low chances of fulfillment, is therefore one of the most important tasks in the definition of the requirements catalog.

Every product developer should always keep these "5 I-s" in mind and be aware that lack of integration during product development is a major cause for their occurrence. In this context, besides the implementation of an IPD, especially tools for a systematic definition of the requirements catalog are of importance. These are explained below.

3.4 Tools for the Identification and Specification of Requirements

Collecting and specifying the requirements for a component is not trivial, which partially results from the wide range of possible requirements that often by far exceed what a single, specialized engineer can have in mind. According to Barg [6] a product design must be

- cost-oriented,
- functional,
- suitable for production,
- suitable for assembly,
- reproducible,
- suitable in terms of quality,
- recyclable/disposable,
- suitable for maintenance,
- and environment-friendly.

The question arises, which further requirements result from the individual development targets. The customer will hand over the answers to the questions that they assume to be relevant, e.g. in the form of a specification sheet. These answers must then be transferred into a requirements catalog, which can include a further precisioning by adding quantitative and qualitative specifications. For this purpose, a "guideline with list of main characteristics" can be used. As already explained, there is a risk of incompleteness concerning the information handed over by the customer. Therefore, the "scenario technique" can be applied to complete the requirements catalog. In order to achieve a solution-neutral formulation of requirements, these should be defined as functionalities to be provided. The tools mentioned above and the procedure for the determination of the functionalities are explained in the following references [1, 4, 7].

3.4.1 Guideline with List of Main Characteristics

In a guideline with a list of main characteristics, the requirements defined by the customer are sorted and specified according to various main characteristics. This is meant to bring up associations, which then provide further insights into the relevant points that in turn can be transformed into further requirements [4].

The advantage of this method is that the systematic representation is not only used to reduce the risk of an incomplete list. Furthermore, it simplifies the work of the design team, as it can be quite easily transferred into a systematic requirements catalog. If the same basic structure is retained for different development tasks, the recurring patterns can also help to reduce the potential for errors. Table 3.1 shows a guideline with a list of main characteristics including some general examples.

This list can now be completed by the scenario technique.

3.4.2 Scenario Technique

The scenario technique is of particular interest in the context of an IPD, because here, at first, the entire product life cycle, including all steps from production to disposal, is considered. For each stage of the product life cycle a scenario is developed and then it is questioned what can happen to the product (e.g. where could it be used) and how it should react. From the answers, requirements can be derived [4]. To derive the requirements, a high level of detail is usually necessary. For this, Kramer [8] presents a three-step procedure in which the customer's statement (1. stage) undergoes a deepening (2. stage) and a clarification (3. stage) in order to derive a usable requirement for the product.

The procedure of the scenario technique is explained using an example:

Bicycles with a CFRP-frame are gaining popularity. A design company specializing in FRP design methods is given the task to design a CFRP bicycle by a bicycle manufacturer without FRP experience. The design company uses the scenario technique to ensure that no critical requirements are missed. After the division of the product life cycle, the section "Use" is being discussed. In response to the question "What can happen with the bicycle in the use phase" the customer replies "transport in a car" and the follow-up question about the desired reaction

Table 3.1 Guideline with list of main characteristics [4][1]

Main characteristic	Examples
Geometry	Size, height, length, diameter, required space, number, arrangement, connection, expansion and extension
Kinematics	Type of movement, direction of movement, velocity, acceleration
Forces	Force size, force direction, force frequency, weight, load, deformation, stiffness, spring characteristics, stability, resonances
Energy	Performance, efficiency, loss, friction, ventilation, state variables such as pressure, temperature, humidity, heating, cooling, connection energy, storage, work input, energy conversion
Material	Physical and chemical properties of the input and output product, auxiliary materials, prescribed materials (food law, etc.), material flow and transport
Signal	Input and output signals, display mode, operating and monitoring devices, signal form
Security	Direct safety technology, protective systems, operational, occupational and environmental safety
Ergonomics	Human–machine interaction: operation, mode of operation, clarity, lighting, design
Manufacturing	Restriction by production site, largest producible dimension, preferred manufacturing process, means of production, possible quality and tolerances
Control	Measurement and testing facilities, special regulations (TÜV, ASME, DIN, ISO)
Assembly	Special mounting instructions, assembly, installation, on-site assembly, foundation
Transport	Limitation by lifting equipment, railway profile, transport routes according to size and weight, mode and conditions of dispatch
Usage	Low noise, wear rate, application and sales area, location of use (e.g. sulfurous atmosphere, tropics …)
Maintenance	Maintenance-free or frequency and time required for maintenance, inspection, replacement and repair, painting, cleaning
Recycling	Reuse, recycling, disposal, storage
Costs	Maximum permissible manufacturing costs, tool costs, investment and amortization
Time plan	End of development, network plan for intermediate steps, delivery time

of the bicycle is answered with "transport in a motor vehicle should be unproblematic." To convert this unspecific statement into a requirement the three-step procedure according to Kramer [8] is applied:

Step 1 (statement)
   Customer request "unproblematic transport in car"

---

[1]Printed with permission of Springer Nature.

Step 2 (detailing)

The customer might have meant:

- *Quick and easy disassembly (without tools) of individual parts for quick reduction of the required transport space*
- *User can store bicycle in the vehicle even at elevated temperatures in summer (e.g. in a parked car)*
- *Compatibility with common transport systems (e.g. roof rack)*
- *[...]*

Step 3 (clarification)

- *Wheels and seat tube are mounted to the frame/fork* via *tool-free quick-clamping systems*
- *Up to 10 h at a temperature of 90 °C are withstood without damage or impairments to the functionality*
- *Hub width front between 70 and 100 mm and rear between 110 and 135 mm*
- *[...]*

The requirements derived in this way by the design office are in turn discussed with the customer and finally incorporated into the requirements catalog.

Using the scenario technique, can reduce the danger of an incomplete requirements catalog. Eventually, the requirements derived from the scenario technique are incorporated into the guideline in the same way as the requirements formulated by the customer right at the beginning.

3.4.3   Determination of Functions and Functional Structures

When specifying the requirements for the product, care must be taken to ensure that these are formulated as solution-neutral as possible. Although the requirements catalog should be complete and the individual requirements should be specific, the design freedom necessary for a high-quality technical solution should not be unnecessarily limited. A proven approach to this is the definition of requirements as functions that are to be fulfilled. Hence, a negative example would be to require:

"To enable fixation of the seat post to the saddle tube using a seat clamp, the single-piece, tubular seat post must have a cylindrical section of at least 150 mm in length with a diameter of 25.4 mm, on the side facing away from the saddle bracket." This requirement already strongly pre-defines the solution. The design shown in Fig. 3.4 would be impossible, even though the two-piece design offers very interesting spring characteristics. Therefore, it would have been better to demand the fulfillment of the function "transfer of the loads acting on the saddle to the seat tube via a saddle clamp."

A systematic method for the determination of the functions to be provided by a component, as well as the interdependencies between the individual functions (functional structure), is also very well suited to ensure that the requirements catalog is complete. Following [9], a possible procedure is presented below:

Fig. 3.4 Split comfort seat post. Images adapted, printed with permission of Canyon Bicycles GmbH

In a first step, the product is described fundamentally. This includes the identification of all elements with which the product can interact during its life cycle. These elements can be, for example, users, environmental influences and other components or systems. In the example of the seat post, for example, the saddle clamp on the seat tube of the bicycle frame would be a relevant element. This consideration is carried out individually for each stage of the product life cycle.

The second step consists of analyzing all possible interactions between the product and the surrounding elements. Based thereon, function that the component must fulfill are derived. Team members from design, manufacturing and material area participate in this step. For example, the interaction of seat tube, seat clamp and seat post results in the function "withstand the clamping forces." For simpler products, it may be sufficient to carry out these first two steps, but for more complex products, it makes sense to link the individual functions to a functional structure in a third step [7]. This way, functions can be divided, for example, into main functions that result directly from the purpose of the product and sub-functions that are necessary to enable or maintain usability.

In a fourth step, quantifiable criteria are defined for each of the individual functions. For the function "withstand the clamping forces," for example, the actual clamping forces can be quantified.

Finally, the criteria are prioritized in a final, fifth step, i.e. for example into demands and wishes.

3.5   Guidelines and Requirements Catalogs for FRPs

Taking into account the findings of the previous sections, the following section presents guidelines for the development of FRP components, intending to assist the practical application of the procedures defined in this book. According to the division of the design team in the three areas "design," "production" and "materials" (as defined in Sect. 2.1), each team member initially focuses on the information required for its specific area. Therefore, a separate guideline is proposed for each area. If the structure of the team differs, however, the information contained in these guidelines must be distributed accordingly. The proposed guidelines have been developed for a single FRP component, which may be integrated into a superordinate structure or contain add-on components. Following the presentation of the individual guidelines, it is shown how they can be combined in an overall requirements catalog.

3.5.1   Guideline for the Design Department

Table 3.2 shows the guideline that should be used by the "design" department in order to sort, specify and complete the customer's requirements for the FRP component. The main characteristics are listed in the left column. The right column shows the information required to carry out the product development.

Since the team member from the design area is also the team leader in the approach proposed in this book (see Sect. 2.1), this guideline also contains information of general interest, for example on topics such as recycling and maintenance. It is the task of the design team member to collect the information below during the discussion with the customer and thereafter respectively. Even if there are no requirements or wishes for a specific feature, this should be recorded so that design freedoms are clearly visible.

3.5.2   Guideline for the Manufacturing Department

The team member from the "manufacturing" department must ensure that all the information needed for the respective development task are given. This results in redundancies to the information already collected by the "design" department. These different perspectives of the departments should be exploited in order to comprehensively define the development task. In addition, further information is required for the manufacturing-related development tasks, which are listed in Table 3.3.

3.5.3   Guideline for the Materials Department

In addition to the information already collected by the other team members, the team member from the materials department needs the information listed in Table 3.4 to fulfill the corresponding development tasks.

Table 3.2 FRP-related guideline with list of main characteristics for the "design" team

Main characteristic	Relevant information
Function	• Basic functionality the part is supposed to fulfill
Part geometry and construction space	• Requirements resulting from functionality (e.g. seating area for seat shell) • Available construction space (maximum thickness) • Position of connection points for integration in superordinate structures • Position of connection points to subordinate structures • Other design restrictions (e.g. resulting from corporate design)
Loads	• Loads: direction, size, introduction point, frequency, duration • Reaction to loads: allowable deformation, service life, fail-safe requirements • Spring/damping properties
Kinematics	• Type and direction of movement • Velocity/acceleration • Frequency and duration
Weight	• Weight target (e.g. minimization or smaller reference value) • Permissible total weight
Time plan	• Time plan for the development (e.g. dates for presentation of concepts, drafts and final recommendations for actions)
Crash	• Energy absorption • Failure behavior (e.g. splintering)
Transport	• Permissible size/geometry of single parts (e.g. resulting from pre-defined transport in shipping container)
Recycling	• Recycling/disposal requirements (e.g. company policy regarding component recycling)
Maintenance	• (Dis-)assembly • Exchangeability of complete part • Exchangeability of single elements • Damage diagnostics • Maintenance cycles, freedom of maintenance
Strategy	• Acceptable development effort • Permissible development risk (e.g. permissible risk that certain technical goals are not achieved) • Specifications regarding the usability of design know-how to be built up for further products • Required flexibility regarding possible changes of the boundary conditions
Legal boundary conditions	• Legal requirements (e.g. resulting from recycling management act)

This provides the information necessary to fulfill the development tasks. This information is then converted into requirements (step 1-2 in Fig. 2.1) and must be combined internally into a requirements catalog (step 1-3 in Fig. 2.1), which is binding for all three departments.

Table 3.3 FRP-related guideline with list of main characteristics for the "manufacturing" team

Main characteristic	Relevant information
Series size	• Short, intermediate and long-term plans for serial size (e.g. series size at product launch and 5, 10 as well as 15 years later)
Costs	• Maximum invest • Maximum part costs • Amortization
Time plan	• Acceptable time to start of production
Usage of resources	• Already available production systems (e.g. for the production of other products) • Existing supplier structure • Available space and possibilities for expansion • Labor capacity and expertise, possibilities for expansion
Processes	• Preferred manufacturing technologies
Quality	• Tolerances • Quality assurances • Surface waviness and roughness • Laminate quality
Post-processing	• Color requirements • Preparation of joining contact areas • Edge trimming • Packaging
Integration in overall structure	• Specifications with regard to joining methods for target component with superordinate structure (e.g. connection via screw bolt) • Specifications with regard to joining methods for target components with subordinate structures (e.g. connection of add-on parts via bonding)
Strategy	• Permissible development risk (e.g. permissible risk that process concept is actually feasible) • Specifications regarding the usability of design know-how to be built up for further products • Required flexibility regarding possible changes of boundary conditions

3.5.4 Overall Requirements Catalog

To create the overall requirements catalog from the individual guidelines of the departments, the departments must

- convert the information into quantified or qualified requirements,
- separate the requirements into demands and wishes and
- prioritize the wishes.

The information on the separation into demands and wishes should already be included in the guidelines.

Table 3.4 FRP-related guideline with list of main characteristics for the "materials" team

Main characteristic	Relevant information
Materials	• Preferred materials and semi-finished products
Climatic boundary conditions	• Sites of operation and transport • Temperature ranges of operation including frequency and duration • Air humidity • Permissible deformation (e.g. resulting thermal expansion or swelling due to humidity)
Corrosive boundary conditions	• Joining contacts (e.g. contact to super-/subordinate structures) • Ambient conditions including frequency and duration (e.g. salt water contact)
Tribologic boundary conditions	• Relative movements to surface (e.g. in slide bearings): Frequency, type/direction of movement, forces
Optical boundary conditions	• Appearance: Transparency, color, shrinking behavior, scratch resistance, surface structure
Strategy	• Permissible development risk (e.g. permissible risk that material concept is actually feasible) • Specifications regarding the usability of design know-how to be built up for further products • Required flexibility regarding possible changes of boundary conditions

Subsequently, it is the task of the team leader ("design" area) to collect and structure the requirements of all departments in a complete requirements catalog. The requirements catalog can then be structured according to different approaches. When working with a guideline with a list of main characteristics, it is advisable to maintain this order. An example for a total requirements catalog is given below. It considers the example of a wall-mountable seating space used as an additional auxiliary seat in a sports boat.[2] Figure 3.5 shows a possible implementation. Table 3.5 shows the overall requirements catalog for the component.

With this overall requirements catalog, milestone 1 of the procedure proposed in Fig. 2.1 is fulfilled.

---

[2]The requirements mentioned serve to illustrate the topic and have been simplified or shortened in this sense—a direct transferability for an actual component is not given.

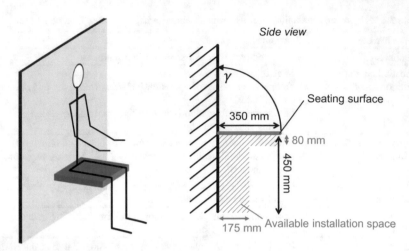

Fig. 3.5 Boundary conditions for wall-mountable FRP seat area

Table 3.5 Overall requirements catalog for a wall-mountable FRP seat area

Main characteristic	Relevant informations
Function	The part is intended to provide a temporary seat for adult persons
Part geometry and construction space	• A square seat area with an edge length of 350 mm ($\pm$10 mm) is to be realized (seating area refers to the surface with direct contact to the passenger) • The seating surface should be mounted on a wall and be plane-parallel to the floor positioned at a right angle relative to the wall; in unloaded condition a slight upward inclination of max. 5° is permissible (see drawing, $85° \leq y_{unloaded} \leq 90°$) • The distance between the seat and the floor must be 450 mm ($\pm$10 mm) • There is no permissible installation space above the seat • No elements of the construction may protrude from the imaginary cuboid that results when the seat surface is extruded toward the floor • In the first half of the seat surface (starting from the wall) the entire installation space between the seat surface and the floor is available (see drawing), in the second half there is a maximum installation space of 80 mm from the seat surface toward the floor • The danger of impact injuries at the two corners facing away from the wall must be avoided, e.g. by rounded edges • Generally, all edges with the risk of impact damage must be secured accordingly, e.g. by rounded edges • A connection to the wall can be made within the specified installation space; the wall is accessible from both sides and consists of a GFRP sandwich structure with a thickness of 15 mm; there are no relevant installation space restrictions on the rear side

(continued)

Table 3.5 (continued)

Main characteristic	Relevant informations
Loads	The following static load cases are to be considered in the design • Load case 1 (corresponds to bending due to static sitting on the front edge): load of 800 N on the front (facing away from the wall) upper edge; the inclination of the seat surface must not exceed $y_{\text{Load case } 1} = 95°$ relative to the wall (see drawing) • Load case 2 (corresponds to torsion due to static sitting on a corner): Point load of 800 N on one of the two front corners; the difference in height between the two front corners must not exceed 15 mm With regard to service life, the following must also be observed • Load case 3: No failure up to 10^4 load cycles of pulsating load (line, 0–1000 N) on the front edge
Crash	• No sharp edges in case of failure
Kinematics	• Not applicable (no folding mechanism)
Weight	• The maximum weight for the component including the connection technology required for fixing to the wall is 3 kg
Integration in overall structure	• The connection to the wall must be designed to be detachable • Drill holes in the wall are permissible up to a maximum diameter of 15 mm, with a required center distance of at least 100 mm
Materials	• Skin-friendly
Climatic boundary conditions	• Continuous use in sea air at ambient temperatures from—10 to 50 °C • Even tropical climatic conditions with high humidity must be permanently tolerated
Corrosive boundary conditions	• Occasional salt water contact (splash water) • Compatibility to common cleaning agents
Tribologic boundary conditions	n/a
Optical boundary conditions	• The surface should have a high degree of gloss; the use of high-quality materials should be visible (wish: "carbon look" with fabric reinforcement) • The seat surface should have a high scratch resistance
Time plan	• 2 months after commissioning: Design Review 1 • 3 months after commissioning: Design Review 2 and presentation of the process concept; determination of the drafts to be elaborated • 5 months after commissioning: presentation of the elaborated design and economic evaluation • 8 months after commissioning: presentation of the results of the prototype tests and the overall techno-economic evaluation; Preparation of a decision on further implementation • Planned start of production: 18 months after commissioning

(continued)

Table 3.5 (continued)

Main characteristic	Relevant informations
Serial size	The following target figures must be considered: • Number of units to be produced in year 1 and 2 after sales start: 100 p.a. • Number of units to be produced in year 3–5 after sales start: 500 p.a. • Number of units to be produced in the following years: 1000 p.a.
Costs	• Maximum invest: 500,000 € • Maximum component costs: 500 € per piece
Usage of resources	• Only suppliers from the European Union
Processes	• Wish: use of existing expertise in vacuum infusion processes or prepreg autoclave technology
Quality	• The tolerance class "medium" of ISO 2768-1 applies to all dimensions on the component • Quality assurance should be carried out solely by visual inspection
Post-processing	• Open component edges (e.g. of a trimming) shall be protected against water penetration by sealing, if water contact might occur
Maintenance	• The component should be exchangeable as a whole • Wish: a relevant threat to structural integrity should be detectable by visual inspection
Strategy	• The risk that the weight targets are to be exceeded by more than 10%, while all other requirements are met, is less than 20%. • In the future, a double-seat variant is also to be developed; usage of the same materials and technologies as for the single-seat variant is intended
Legal boundary conditions	• None
Transport	• No restriction
Recycling	• No hazardous waste

3.6 Questions for Self-Check

Below are some questions and tasks to help you reflect on the main contents of this section. The solutions can be found in Chap. 7.

> R7. Give two examples for tools that can help in the process of creating a requirements list.
> R8. Name four risks when creating a requirements list.
> R9. For each of the areas "design," "manufacturing," and "materials," name three relevant pieces of information that a requirements list should contain.

R10. Name two types of demands and wishes and explain the differences.
R11. Explain the terms fixed demand and interval demand.

Literature

1. Ehrlenspiel, K., Meerkamm, H.: Integrierte Produktentwicklung: Denkabläufe, Methodeneinsatz Zusammenarbeit. Carl Hanser Verlag GmbH Co KG, Munich (2013)
2. Lindemann, U.: Vorlesung: Methoden der Produktentwicklung, Teil 1., Technische Universität Munich (2001)
3. Naefe, P.: Einführung in das Methodische Konstruieren: Für Studium und Praxis. Springer Fachmedien Wiesbaden (2012)
4. Pahl, G., Beitz, W., Schulz, H.-J., Jarecki, U.: Pahl/Beitz Konstruktionslehre: Grundlagen erfolgreicher Produktentwicklung Methoden und Anwendung. Springer-Verlag, Berlin/ Heidelberg (2013)
5. Lindemann, U.: Methodische Entwicklung technischer Produkte: Methoden flexibel und situationsgerecht anwenden. Springer, Berlin (2006)
6. Barg, A.: Recyclinggerechte Produkt- und Produktionsplanung. VDI-Z Integrierte Produktion **133**(11) (1991)
7. Verein Deutscher Ingenieure e.V.: VDI 2221—Methodik zum Entwickeln und Konstruieren technischer Systeme und Produkte (1993)
8. Kramer, F., Kramer, M.: Bausteine der Unternehmensführung: Kundenzufriedenheit und Unternehmenserfolg. Springer-Verlag, Berlin/Heidelberg (2013)
9. Breuer, U.P.: Commercial Aircraft Composite Technology. Springer, Berlin (2016)

Chapter 4
Phase 2: Conception/Drafting Phase

Abstract Based on the requirements catalog, conception and drafting of solutions is carried out. **Conception** is the development of a basic solution (→the concept), and **drafting** is the development of a quantitative solution (→the draft). Compared to the conventional product development, an IPD intends to achieve an improved coordination through strong parallelization. Thereby, efficient working requires that the fundamentals for the developments in the various departments are created comparatively quickly. Thus, the design conception is initially based on rough calculations, so that, for example, the basis for the process design is available at an early stage. In this section, at first, fundamentals for the design with FRPs are explained. The following sections deal with the further development and show, step for step, how to deal with:

- concept development and pre-dimensioning
- selection of fiber material and structure of the fiber reinforcement
- development of a process concept
- decision for thermoset or thermoplastic
- revision and creation of an overall draft.

4.1 Overview

Figure 4.1 shows the procedure of phase 2, for the creation of concept and draft.

4.2 Fundamentals of Product Design with FRPs

Like any other material group, FRPs have both advantages and disadvantages, which must be taken into account in product design, in order to achieve the best possible result. Issues to be taken into account mainly result from the heterogeneous fiber-matrix structure. Ignoring these issues causes a deficient exploitation of the material potential, while on the other hand a profound knowledge offers the opportunity to exploit the enormous potential that is offered by the fiber-matrix

© The Author(s), under exclusive license to Springer Nature Switzerland AG 2021   49
D. May, *Integrated Product Development with Fiber-Reinforced Polymers*,
Engineering Materials, https://doi.org/10.1007/978-3-030-73407-7_4

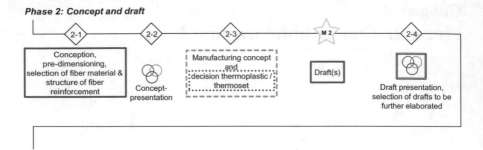

Fig. 4.1 Overview of phase 2—conception and drafting

structure. That is why this sub-section aims to provide the basic material—expertise that is required for a successful FRP product development.

4.2.1 Relevance of the Fiber Volume Content

With regard to the effectivity of the fiber reinforcement of polymers, the **critical fiber volume content** FVC_{crit} must be considered. As shown in Fig. 4.2, below FVC_{crit} incorporation of fibers weakens the polymer mechanically, whereas above FVC_{crit} a reinforcement takes place. As long as FVC_{min} is not exceeded, the stress on the matrix at failure of the fibers σ'_m is smaller than the tensile strength of the remaining matrix $R_{m,M}$. This means that the composite strength equals the tensile strength of the remaining matrix, which is smaller than the strength of the neat polymer by the factor FVC (fiber volume content), as the fibers reduce the useable cross section of the matrix. Above FVC_{min} failure of the composite corresponds to the failure of the fibers. Yet, only when FVC_{crit} is exceeded an increasing exploitation of the fiber tensile strength $R_{m,F}$ leads to a mechanical improvement compared to the neat polymer [1, 2].

Accordingly, FVC_{crit} can be defined as:

$$FVC_{crit} = \frac{R_{m,M} - \sigma'_m}{R_{m,F} - \sigma'_m} \tag{4.1}$$

For a classic fiber-matrix combination, such as glass fiber-reinforced polypropylene, FVC_{crit} is e.g. below 10% [3, 4] and thus significantly below the fiber volume contents normally found in technical applications.

A theoretical upper limit for the fiber volume content is given by the maximum packing density (about 90.7% for circular cross-sections in hexagonal packing, i.e. perfectly parallel arrangement). Practically, the upper limit for structures with intended parallel fiber orientation is at about 80% due to imperfect fiber alignment and distribution. An additional limitation is given by the fact that a complete

Fig. 4.2 Illustration of the critical fiber volume content Adapted from [1] (Image adapted, printed with permission of Taylor & Francis)

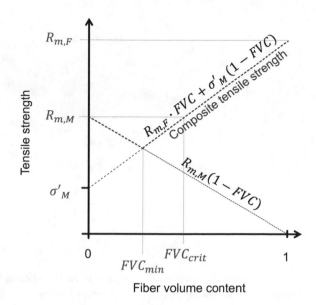

wetting of all individual fibers must always be guaranteed and must be realizable in terms of process engineering. Hence, a process- and material-specific optimum of the fiber volume content exists. Therefore, even in structures with intended completely parallel fiber orientation, a fiber volume content of more than 70% is rarely aimed for. Due to the specific arrangement of the fibers, fabric-reinforcements typically provide fiber volume contents from 50 to 60% and random fiber mats from 20 to 40%. [5, 6]

4.2.2   Relevance of Fiber Length and Orientation

When looking at Fig. 1.3, it becomes clear that the length of the fibers and their orientation relative to the applied load play a significant role for the mechanics of FRPs. Figure 4.3 schematically illustrates the relevant variables for the introduction of a tensile force into the fibers within a short or long fiber-reinforced polymer. On the one hand, the load can be introduced via the front surface area A_{F1}, a mechanism that, however, is usually negligible. The largest part of the load introduction occurs via shear forces on the outer surface A_{F2}.

The stress in the fiber σ_F thus results from the force in relation to the front surface plus the shear force τ related to the outer surface (in the case of continuously fiber-reinforced plastics, the load is completely introduced via shear forces, so the corresponding term for force introduction into the front face is zero):

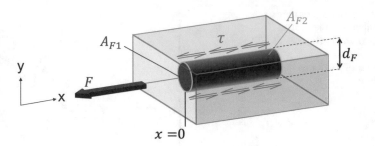

Fig. 4.3 Load introduction into the fibers of a short/long fiber-reinforced polymer

$$\sigma_F = \frac{F_{(x=0)}}{A_{F1}} + \frac{\pi d_F}{A_{F2}} \int_0^x \tau \, dx \qquad (4.2)$$

This formula can also be used to derive the critical fiber length l_{crit}. Given an interfacial shear strength τ_{max}, the critical fiber length defines the fiber length that is necessary to fully exploit the strength of the fibers. For this, the maximum stress that can be introduced into the fibers, must be greater than the breaking stress $\sigma_{F,break}$ of the fibers

$$l_{crit} = \frac{\sigma_{F,Break} \cdot d_F}{2\tau_{max}} \qquad (4.3)$$

Only if this critical fiber length is exceeded, the fiber properties are fully exploited. Thereby, the exact value depends on both the fiber strength and the adhesion between fiber and matrix. The worse the adhesion, the greater the critical fiber length. In [3], for example, the properties given in Table 4.1 were used to determine the critical fiber length for glass and carbon fiber-reinforced polypropylene, resulting in a value of 0.89 mm and 0.81 mm, respectively. The interfacial shear strength was determined in a single fiber pullout test by normalizing the measured maximum force to the outer surface. The authors themselves state that this method is error-prone, since it assumes that the shear stress is constant along the fiber axis, which is not the case due to the elasticity of the material partner. The same applies to the calculation of the critical fiber length based on this value. These values can therefore only serve as an indicator.

The influence of fiber length varies for different mechanical properties. As Fig. 4.4 shows on the example of a glass fiber-reinforced polypropylene, the maximum stiffness of this FRP is almost reached at only a few millimeters fiber length, while the strength keeps increasing up to a fiber length of about 10 mm. The difference can be explained by the fact that the shear strength of the boundary phase is not fully utilized when measuring stiffness, since this is not measured in the failure area, as it is the case with tensile strength. The figure also shows that the

Table 4.1 Calculation example for the critical fiber length

Fiber material	Strength in MPa	Fiber radius in μm	Interfacial shear strength to PP in MPa	Critical fiber length in mm
Glass fiber (GF)	1956	6.90	15.2	0.89
Carbon fiber (CF)	3950	3.75	18.2	0.81

Fig. 4.4 Influence of the fiber length on the mechanical properties of a FRP. Adapted from [7] (Image adapted, printed with permission of Elsevier B.V.)

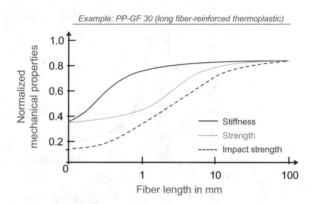

most critical case regarding the fiber length is given by the impact strength. An increasing fiber length results in increasing impact strength up to about 100 mm.

The use of long fibers and above all continuous fibers also offers the great advantage that a more precise orientation can be achieved. As depicted in Fig. 4.5, the relative orientation between the longitudinal axis of the fiber and the applied load is of outstanding importance. Even a few degrees of deviation cause a severe reduction in stiffness and strength. The minimum is reached transverse to the longitudinal axis of the fiber. This is quite easy to understand when considering the structural substitute model of the FRP, which is illustrated in Fig. 4.6:

- If a load is applied to a FRP in longitudinal direction of the fibers, the interaction of fibers and matrix corresponds to a parallel loading of springs. In a first approximation, the stiffness of the composite in this direction ($E_{\text{Composite},\|}$) can therefore be described by a corresponding rule of mixture based on the stiffness of the fiber (E_{Fiber}) and the matrix (E_M) [2]:

$$E_{\text{Composite},\|} = E_{\text{Fiber}} \cdot \text{FVC} + E_M \cdot (1 - \text{FVC}) \tag{4.4}$$

Accordingly, the strain is always the same, whereas the stress is different in fibers and matrix. Due to their relatively high stiffness, the vast majority of the load acts on the fibers. FRPs should be designed to specifically generate such loads.

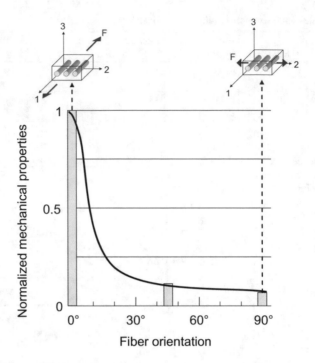

Fig. 4.5 Influence of the fiber orientation on the mechanical properties of a FRP

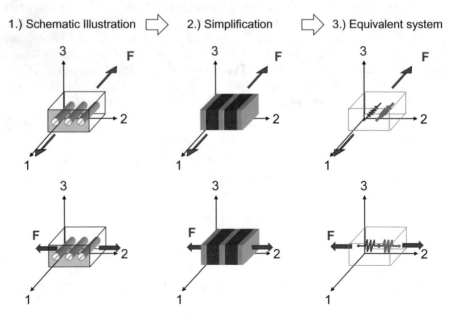

Fig. 4.6 Derivation of substitute systems for elongation of FRPs with unidirectional fiber orientation

- With a load transverse to the fiber, the stress on fibers and matrix is the same, whereas the strain differs. The substitute system therefore corresponds to a serial connection of springs and a first approximation of the directional stiffness of the composite ($E_{Composite,\perp}$) is given as follows [2]:

$$E_{Composite,\perp} = \frac{E_M}{1 + FVC\left(\frac{E_M}{E_{Fiber}} - 1\right)} \tag{4.5}$$

Such a load is extremely unfavorable, as the strength is determined by the matrix strength and the fiber-matrix adhesion.

If the FRP is not considered at the fiber level, but on a higher level—the laminate level—there are further points to be considered, which is explained in detail in the next section.

4.2.3   Laminate Design

Corresponding to the previous section, the maximum lightweight potential of FRPs is reached with a continuous fiber orientation that is adapted to the load situation. For this, often several layers of fibers are combined in a laminate. Three different basic types of individual laminate layers can be distinguished, with regard to the existing fiber orientation (Fig. 4.7):

- **unidirectional** laminate layers (all fibers are oriented in parallel)
- **quasi-isotropic** laminate layers (in an idealized perception, an infinite number of symmetry planes exists, perpendicular to the laminate plane, so that there is no preferred fiber orientation; this is, e.g. the case for random fiber mats with randomly oriented fibers)

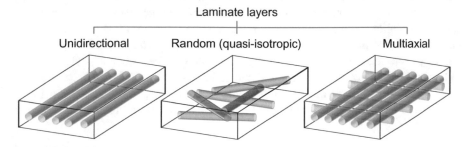

Fig. 4.7 Types of laminate layers

- **multiaxial** laminate layers (e.g. woven or braided fabrics with fibers in two or more specific orientations)

The orientation of the laminate layers in the composite laminate determines its directional stiffness and strength. Due to the strong influence of fiber orientation on the mechanical properties, the lightweight potential of a FRP design depends on how good the possibility for an anisotropic design, adapted to the load situation, was exploited. The example shown in Fig. 4.8 illustrates this, by comparing a steel beam to an aluminum variant and different FRP variants with continuous carbon fiber reinforcements, differing in orientation (see Sect. 4.2.4 for an explanation of the notation).

The beam is clamped on one side (1000 mm free length at 100 mm width and 10 mm height) and is loaded with a weight of 1 kg (under gravity). The upper table shows the resulting deflection for different materials when the geometry is identical. The mass of the aluminum variant is only a third of that of the steel variant. However, the deflection is also significantly greater because the modulus of elasticity is only around a third. Equivalently, the deflection of the quasi-isotropic (CFRP-QI), orthotropic (CFRP-OT) and even the unidirectional (CFRP-UD) CFRP version (all with HT fibers and 55% fiber volume content, see Sect. 4.4 for further explanation) is higher than for the steel beam. Simple substitution without design adaptions therefore usually brings no advantages regarding stiffness. It is key to exploit the material-specific advantages. In the case of aluminum, for example, these would be the design freedoms given for die-cast aluminum. For FRPs, above all these are the possibilities for adaption of the fiber orientation to the load situation. The table in Fig. 4.8 shows the potential of a consistent exploitation of these advantages. If the thickness of the beam is adjusted so that all versions have a deflection of 1.9 mm, then the unidirectionally reinforced variant is almost 80% lighter than the steel variant, at equal functionality.

For the design of a FRP, the stiffness and strength of the complete laminate is relevant. To determine the properties of a laminate composite from the properties of individual laminate layers, the classical laminate theory was developed, which is explained in more detail in Sect. 4.4.3. However, in addition to these values it must be taken into account that the laminate structure can induce mechanical couplings, as the individual layers can show different deformation behavior based on different fiber orientations and positions within the laminate. Figure 4.9 shows this for the example of a simple plate that is reinforced only by 45° fiber layers. If a strain is applied, this leads simultaneously to a shear, due to the different stiffness of the plate in the direction of the fibers and perpendicular to them (Fig. 4.6). Even with a thermal strain, this asymmetrical behavior becomes apparent. This behavior is referred to as a strain-shear coupling. In addition to this, the couplings illustrated in Fig. 4.10 can occur:

(a) strain-shear: a tensile load induces not only elongation but also shearing and vice versa.

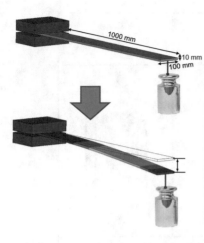

Identical geometry (1000 x 100 x 10 mm³):	Weight	Deflection u
Steel	7.85 kg	5.7 mm
Aluminum	2.68 kg	17.1 mm
CFK-QI [0°/+45°/90°/-45°]$_{nS}$	1.45 kg	20.0 mm
CFK-OT [0°/90°]$_{nS}$	1.45 kg	19.0 mm
CFK-UD [0°]$_n$	1.45 kg	9.5 mm

Identical function (adapted height):	Weight at $u = 1.9$ mm	
Steel	7.85 kg	100.0%
Aluminum	3.87 kg	49.3%
CFK-QI [0°/+45°/90°/-45°]$_{nS}$	2.20 kg	28.1%
CFK-OT [0°/90°]$_{nS}$	2.16 kg	27.5%
CFK-UD [0°]$_n$	1.72 kg	21.9%

Fig. 4.8 Example for the influence of anisotropy exploitation on the lightweight potential (for explanation of the laminate notation see Sect. 4.2.4). Left image adapted, printed with permission of Leibniz-Institut für Verbundwerkstoffe GmbH

(b) strain-bending: a tensile load induces not only elongation but also bending and vice versa.
(c) strain-torsion: a tensile load induces not only elongation but also torsion (twisting) and vice versa.
(d) bending-torsion: a torsion moment induces not only a torsion but also bending and the other way around.
(e) shear-torsion: a shear load induces not only shear but also torsion and vice versa.

In order to counteract unwanted couplings and asymmetrical deformations, laminate structures should always be symmetrical, as shown in Fig. 4.11; i.e. the layer structure should be equal on both sides of the centerline. Also for each layer with a specific fiber orientation, there should be a compensating layer with a fiber orientation correspondingly rotated by 90°. When using unidirectional single layers, however, it is still not possible to eliminate couplings completely, because single layers and their compensation layer never have the exact same position within the laminate. If, for example, the laminate shown in Fig. 4.11 (left) is bent, the −45°— layers are further away from the neutral fiber than the +45°—layers, which means that the layer and the compensation layer are not equally loaded.

As often occurs when applying FRPs in practice, these challenges also offer unique design opportunities. The couplings can also be used in a targeted manner, which is an often underestimated advantage of FRPs. Figure 4.12 shows an example from aircraft design. Forward swept wings are considered to be aerodynamically advantageous, but result in enormous requirements concerning the wing stiffness, in order to prevent unwanted torsion. With a metal design, this results in a

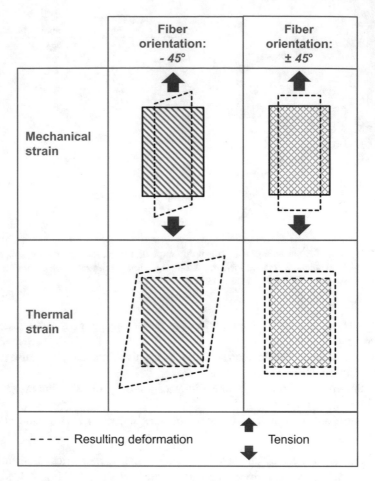

Fig. 4.9 Asymmetrical deformation behavior under mechanical and thermal loading

comparatively high weight compared to backward swept wings. In the "Grumman X-29" aircraft, however, an FRP design with an adapted layer structure implements a bending-torsion coupling, which compensates unwanted torsional deformation under aerodynamic load [9]. The same effect can be used so that rotor blades of wind power plants turn out of the wind if it gets too strong and could damage them [10].

Since FRPs are often used with the aim of weight reduction, often core materials are incorporated in the laminate to form a **sandwich**. This creates a shear-resistant composite of thin-walled, strong and rigid FRP face layers with a thick-walled core. The light core material induces a distance between the highly rigid face layers, resulting in a strong increase of the area moment of inertia at minimal weight increase. Common core materials used in combination with FRPs are foams, honeycomb structures or light woods such as balsa wood.

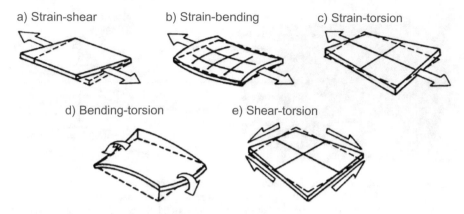

a) Strain-shear b) Strain-bending c) Strain-torsion

d) Bending-torsion e) Shear-torsion

Fig. 4.10 Possible mechanical couplings of FRP laminates. Adapted from [8] (Image adapted, reproduced with the permission of the Verein Deutscher Ingenieure e. V.)

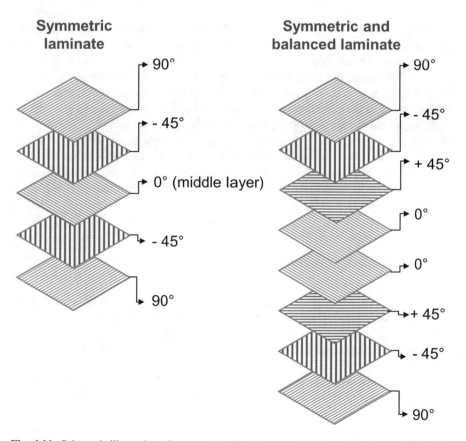

Fig. 4.11 Schematic illustration of a symmetrical and a symmetrical as well as balanced laminate

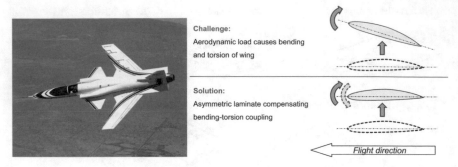

Fig. 4.12 Example for the exploitation of a mechanical coupling to improve the flight characteristics of an aircraft. Left image: This image was catalogued by Armstrong Flight Research Center of the United States National Aeronautics and Space Administration (NASA) under Photo ID: EC87-0182-14. The image is license free because it was created exclusively by NASA. The NASA Copyright Policy states that "NASA material is not copyrighted unless otherwise noted" (see: PD—USGov, NASA Copyright-Guidelines or JPL Image Use Policy)

Finally, another characteristic of textile-reinforced FRPs, resulting from the layered laminate structure, is the danger of delamination. Even if the textiles in the component are pressed together so that no continuous zone of pure resin is visible between the layers, the interface is a weak point in which cracks can propagate without having to pass a layer of fibers. The performance of FRPs under loads perpendicular to the central plane is comparatively low and should be avoided for example by means of supporting elements (Fig. 4.13). In addition, stress peaks between the layers should be minimized, which is why abrupt changes in wall thickness are unfavorable, as they promote the occurrence of cracks in the laminate. A gradual increase leads to a more favorable stress distribution. Hence, the ratio from step width to height should be greater than 10, with individual step heights less than 0.7 mm [11] (Fig. 4.13).

4.2.4   Laminate Notation

For a simple description of laminate structures, specific notations are used. Basically, a laminate is described by writing the respective fiber orientations of the individual layers within square brackets, starting from the lowest layer in positive z-direction (perpendicular to the layer plane, from bottom to top, Fig. 4.14). The orientation angles are separated by a slash. Both, notations with and without degree symbols, are common [12]. Figure 4.14 shows an example. This notation assumes that all layers are of equal thickness, so any deviation from this must be indicated separately. If the laminate is symmetrical, the upper half corresponds to a mirrored variant of the lower one, which can be indicated by a subscript "s" after a square

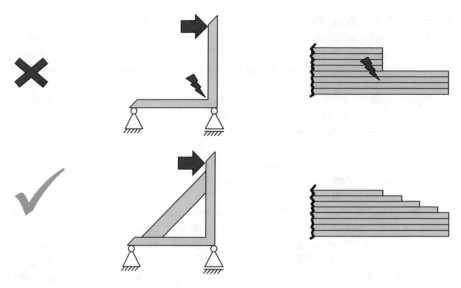

Fig. 4.13 Minimizing the danger of delamination through loading perpendicular to the central plane (left) and thickness changes (right)

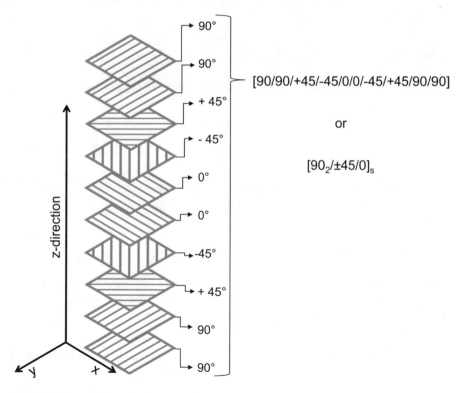

Fig. 4.14 Example for a laminate notation

bracket only containing the lower half of the layers. If two neighboring layers, have identical orientation angles but with different signs, they can be summarized by using the "±" symbol. If several layers of the exact same orientation follow each other, they can be summarized using an index corresponding to the number of layers [12].

Additionally, a subscript number after the square bracket can indicate that the structure defined within the brackets is repeated accordingly. This is often combined with a symmetry index. Accordingly, for example [0/90/0/90/90/0/90/0] can also be noted as $[0/90]_{2S}$.

The VDI guideline 2014 [8] also provides for the possibility of specifying the fiber material or even the type of semi-finished product, e.g. $[0^C/\pm 45^G/90^C]_s$ or $[0_2/\pm 45^{\text{Weave}}/90/\pm 45^{\text{Weave}}/0_2]$.

With regard to short and long fiber-reinforced polymers there are also specific notations to be mentioned. Here commonly an abbreviation for the matrix polymer, followed by the abbreviation for the fiber material and its weight percentage, is used. For example, PP-GF30 for a polypropylene reinforced with 30 wt% glass fibers. Conversion of fiber mass content to fiber volume content can be done with knowledge of the density of material partners according to the following equations. For the conversion of the fiber volume content φ to the fiber mass content ψ, taking into account the matrix density ρ_{Matrix} and the fiber density ρ_{Fiber}, the following applies [13]:

$$\psi = \frac{\varphi \cdot \rho_{\text{Fiber}}}{\varphi \cdot \rho_{\text{Fiber}} + (1 - \varphi) \cdot \rho_{\text{Matrix}}} \qquad (4.6)$$

In addition, the following applies for the conversion of the fiber mass content to fiber volume content:

$$\varphi = \frac{1}{1 + \frac{1-\psi}{\psi} \cdot \frac{\rho_{\text{Fiber}}}{\rho_{\text{Matrix}}}} \qquad (4.7)$$

Both equations assume that air inclusions/pores can be neglected.

4.2.5   Basic Design Methods for FRPs

Due to their excellent specific properties, FRPs are outstanding lightweight design materials. It is therefore of utmost importance to consider them against the background of a holistic lightweight design concept. This requires the consideration of all aspects of lightweight design, as shown in Fig. 4.15:

- **Material lightweight design**: Measures to minimize weight based on usage of materials with relatively high specific properties (such as FRPs).

Fig. 4.15 Lightweight
design methods

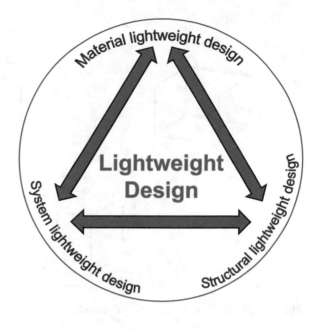

- **Structural lightweight design**: adaptation of the structure to the given load situation and resulting load paths.
- **System lightweight design**: Multifunctionality of components through functional integration beyond the load-bearing function (e.g. load-bearing lightning protection).

The use of FRPs is often limited to material lightweight design. However, the pure substitution of, for example, metal components by structurally unmodified FRP components is rarely successful, because it is necessary to make use of the possibilities for load adaptation of the fiber structure (structural lightweight design) and to take into account the specific characteristics resulting from the fiber-matrix structure (Table 4.3). Due to the heterogeneous structure, also system lightweight design can be achieved by an appropriate selection of fiber and matrix material.

Given this background, the selection of the overall design method is of particular interest when developing an FRP component. Figure 4.16 shows a basic classification of these design methods [14]:

- **Integral design**: The component consists of as few individual parts as possible, each taking over several functions and being geometrically very complex. The part(s) are manufactured in one step from a single material. Typical examples are monocoques for sports cars.
- **Differential design**: The component consists of a comparatively high number of easy-to-manufacture parts of limited geometric complexity and, if necessary, of different materials, which are joined together in further process steps. A typical example is given by bicycle frames made of aluminum, which are welded from several individual tubes.

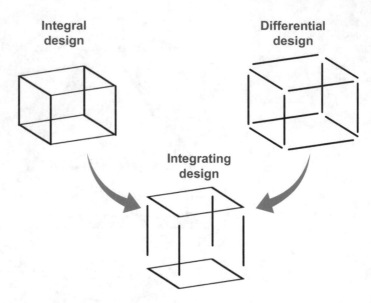

Fig. 4.16 Illustration of different design methods

- **Integrating design**: an attempt to combine the advantages of integral and differential design by targeted evaluation of the advantages and disadvantages
  when dividing the overall structure into individual parts.

In Table 4.2, the potential advantages and disadvantages of the integral and
differential design methods are compared with regard to product development with
FRPs.

It becomes clear that neither the integral nor the differential design is generally
suitable for the product development with FRPs, especially when considering the
overall product life cycle. The goal of IPD with FRPs must therefore be, to combine
the advantages of both design methods in an integrating design. This requires,
among other things, the evaluation of design targets (e.g. homogeneous force flow)
as well as production targets (e.g. simple, robust production), which once again
emphasizes the importance of interdisciplinary cooperation.

The ideal degree of integration is thus dependent on many factors, which are
interrelated and must be balanced. Accordingly, the decision concerning the degree
of integration should take place within the conception phase, as it strongly affects
the complete product life cycle. At the same time, it is of high importance to take
this decision within the framework of an IPD, involving all relevant areas, whereby
the advantages and disadvantages are jointly evaluated.

Table 4.2 Advantages and disadvantages of integral and differential design

Phase	Integral design	Differential design
Design	**Design**	
	+ Homogeneous force flow in single part + Low weight − Part size limited by manufacturing technology	+ Limited complexity of single parts + Flexible adaption (e.g. size adaption) − Additional work for design of joints − Additional weight through joining − Notch effects, stress concentration at joints
	Material selection	
	+ One material must meet all requirements + Local adaption difficult	+ Flexible material selection according to local requirements − Challenges through inhomogeneous thermal expansion
Manufacturing	**Procurement**	
	+ Small number of single parts and materials	+ Simplified usage of standard parts − High number of single parts and eventually materials
	Manufacturing	
	+ Small number of process steps + Low assembly effort + Established manufacturing technologies for FRP available − Complex manufacturing technologies − Selection of manufacturing technologies limited − High manufacturing risk	+ Single processes with limited complexity + Parallelization of work on different parts possible + Process combinations possible + Reduced manufacturing risk + High material efficiency − High number of process steps − High assembly effort
	Tools	
	+ Few tools − Complex tools	+ Simple tools − High number of tools
Distribution/ usage	**Transport**	
	− Adaption of packaging to part required − Challenging transport of large parts	+ Pre-assembly to simplify transport and packaging − Eventually final assembly on-site required
	Maintenance/repair	
	− Very unfavorable damage behavior, as no local replacement is possible − Very few repair methods suitable for FRPs established	+ Exchange of single parts possible

(continued)

Table 4.2 (continued)

Phase	Integral design	Differential design
End of life	**Disposal/Recycling**	
	+ Uniform material usage simplifies disposal[a] − For large parts cutting required before recycling/disposal	− Multi-material design requires dismounting to separate materials[a]

[a]Especially in the field of composite materials, these advantages and disadvantages of the design methods are not necessarily given. For example, an integral design method may also require a material separation if a composite material is used whose material partners do not offer a common recycling option (see also Sect. 5.4.4). On the other hand, there can also be a joint recycling possibility of the material partners for a multi-material design, e.g. ,for composites of FRP and compatible purepolymer

4.2.6 Advantages and Disadvantages of FRPs

Table 4.3 summarizes the most important advantages and disadvantages that should be considered when designing with FRPs. See also Sect. 1.1.2.

The **aim of IPD with FRP** is, through interdisciplinary cooperation,

- to exploit material-specific advantages and
- to compensate/minimize material-specific disadvantages,

taking into account the entire product life cycle.

4.3　Definition of Critical Load Cases and Derivation of Requirements on Geometry and Material

In a first step, the design department needs to create a basic concept for the product, based on the requirements catalog. Thereby, the specific characteristics of FRPs that have been described in the previous sections must be taken into account. Besides this, conventional design methods and tools are used, as they are, e.g. described in the VDI Guideline 2221 [15] or [16, 17]. These design methods will not be considered in detail in this book. However, an example will be used to illustrate how simple calculations can quickly create the basis required for simultaneous development in the different departments. This example is partially based on the results of a research project of IVW GmbH in cooperation with Pfaff GmbH[1] (further details can be found in [18]). However, the calculations and explanations have been adjusted and simplified, respectively, in order to improve the understandability of the procedure.

---

[1]Project "Angepasste Faser-Kunststoff-Verbunde durch verfahrensintegrierte Eigenschafts cbeeinflussung," funded by the German federal ministry of research and education (03NN3113C).

Table 4.3 Typical advantages and disadvantages of FRPs

Advantages	Disadvantages
+ Low density	− Low modulus of elasticity compared to
+ High specific stiffness and strength	steel
+ Targeted fiber orientation according to load	− Often relatively high material costs
situation	− Lack of standardized characterization
+ Tailored deformation behavior	methods for material properties
+ Targeted use of mechanical couplings	− Costly determination of material
+ Wide range of properties due to large variety	properties
of materials and material combinations	− Long-term behavior partly problematic
+ Numerous manufacturing processes for	(aging, creeping)
different materials, component geometries and	− Difficult post-processing
series sizes	− Costly quality assurance
+ Great freedom of design	− Repair relatively difficult
+ Polymer matrix enables corrosion resistance,	− Recycling usually causes loss of
electrical/thermal insulation	properties (downcycling)
+ High degree of integration possible	− In practice, frequent lack of experience
+ X-ray transparency	and expertise cause incomplete exploitation
+ High energy absorption in case of crash	of potential

The focus of the example is on a thread lever (Fig. 4.17). In lockstitch sewing machines, when the needle goes down the thread lever feeds the thread (upper thread), so that it can be wound in a loop around the thread spool in the sewing machine foot. Then this loop must be pulled back by the thread lever to create the knot from the upper and lower thread. Hence, at every stitch the thread lever makes an up and down movement, leading to up to 5000 cycles per minute, at accelerations of up to 25,000 m/s^2. The project objective was to substitute the aluminum thread lever (die-cast aluminum) by an appropriate FRP component. The intention was to reduce the weight and thereby enable usage of smaller bearings so that eventually an increased number of gear revolutions per minute could be achieved.

First, the critical load cases are identified, considering the geometric boundary conditions. Figure 4.18 shows the defined positions for the two bearings, needed for the connection to the superordinate structure, as well as the eyelet for the guidance of the sewing thread. These geometries are marked with green dashed lines in the figure and are predetermined both in their diameter and in their relative position to each other. Accordingly, the component must connect the bearings and the eyelet and transfer the forces between them. Based on a sensor-supported recording of the bearing forces during sewing, the maximum forces can be identified. Figure 4.18 shows the two critical load cases:

- Load case 1: Quasi-static tensile load on the cross section between bearing 1 and bearing 2.
- Load case 2: Quasi-static bending load on the cross section between bearing 2 and the eyelet.

In the next step, these critical load cases and the component areas affected by them are abstracted into static equivalent systems and it is defined how the

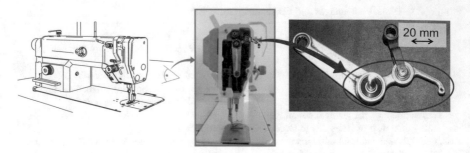

Fig. 4.17 Sketch and front view of a sewing machine and detailed view on a thread lever. Images adapted, printed with permission of Leibniz-Institut für Verbundwerkstoffe GmbH

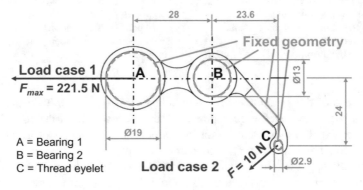

Fig. 4.18 Load cases for the thread lever and pre-defined geometrical characteristics. Image adapted, printed with permission of Leibniz-Institut für Verbundwerkstoffe GmbH

component may react to the load. This is shown in Fig. 4.19. The component area between bearing 1 and 2 may show an elongation due to the tensile load F_{max} (load case 1) $u_{x,max}$ of not more than 5 μm. The bending moment on the component area between bearing 2 and the thread eyelet (load case 2), resulting from the force on the thread eyelet, may cause a bending deformation $u_{x,max}$ of not more than 40 μm.

In a simple manual calculation, the deformation equations necessary for further consideration can be set up. For load case 1 applies:

$$u_x = \frac{F_{max} \cdot L_1}{E_x \cdot A} = \frac{221.5 \text{ N} \cdot 28 \text{ mm}}{E_x \cdot A} < 0.005 \text{ mm} \tag{4.8}$$

Here L_1 is the length of the area between the two bearings, E_x is the stiffness of the material used there in this direction and A is the minimum cross-sectional area (the weakest point is decisive). From this is follows:

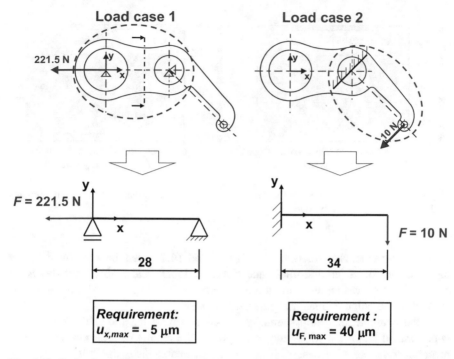

Fig. 4.19 Derivation of equivalent static substitute loads for the critical load cases. Images adapted, printed with permission of Leibniz-Institut für Verbundwerkstoffe GmbH

$$E_x \cdot A > 1.26 \times 10^6 \left[\frac{\text{N}}{\text{mm}^2} \cdot \text{mm}^2 \right]. \tag{4.9}$$

The following applies for load case 2[2]:

$$u_F = \frac{F \cdot L_2^3}{3 \cdot E_x \cdot I} = \frac{10\,\text{N} \cdot (34\,\text{mm})^3}{3 \cdot E_x \cdot I} < 0.04\,\text{mm}. \tag{4.10}$$

Here L_2 is the length of the area between bearing 2 and the thread eyelet and I is the area moment of inertia. It follows:

$$E_x \cdot I > 3.28 \times 10^6 \left[\frac{\text{N}}{\text{mm}^2} \cdot \text{mm}^4 \right] \tag{4.11}$$

---

[2]This equation applies to the case of a constant cross-section, which does not correspond to the optimal design. For a first manual calculation, this assumption is nevertheless useful, as it provides additional certainty that the finally selected combination of fiber material and structure of the reinforcement shows sufficient mechanical performance.

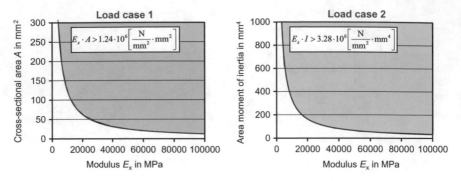

Fig. 4.20 Minimum requirements for the combination of cross-sectional area and Young's modulus of the material to be selected

The requirements defined by Eqs. (4.9) and (4.11) can be met by different combinations of material stiffness and cross-sectional area. Figure 4.20 shows, in red and blue, curves representing the minimum that must be met. All combinations located in the colored areas are generally suitable.

In this example, the component is stiffness-driven; i.e. the allowable deformation is the critical factor, on which the derivation of the requirements for geometry and mechanical properties is based. However, in other cases the strength is in the focus. In this case, it must be checked which maximum loads occur at which point. The basic procedure remains the same. If there is uncertainty whether a stiffness- or a strength-driven design should be carried out, the design can first be based on strength. In any case, for a strength-driven design, the solution should be checked with regard to the permissible deformation and, for a stiffness-driven design, with regard to the permissible stresses. This check can be performed within the scope of pre-dimensioning (see Sect. 4.5).

The described procedure shows how the basic requirements for the component, regarding the mechanical load cases, can be transferred into requirements for geometry and material. This provides a basis for initial, fundamental considerations concerning the material selection.

4.4 Definition of Fiber Material and Structure of the Fiber Reinforcement

At this point, the design department has to make initial decisions regarding the materials to be used. This then forms the basis for the pre-dimensioning and the development of the process concept. It must be decided,

- which **fiber material** and
- which **reinforcement structure** is to be used.

4.4.1 Fiber Materials

The market for fibers is characterized by a very large number of variants, but considering the uncertainties given at this stage, for example with regard to the geometry, at this point only the decision for a certain material group is required. As a basis for this decision, in the following, the most important fiber materials are briefly introduced. This is followed by a comparison based on the most important properties. Figure 4.21 shows an overview of fiber materials relevant for the production of FRPs.

In industrial terms, the glass fibers are by far the most important fibers, followed by carbon fibers. Industrial usage of other fibers lags far behind.

In Tables 4.4, 4.5, 4.6, 4.7, 4.8, 4.9, 4.10 and 4.11 selected fiber materials are presented in more detail.

Table 4.12 shows a comparison of the fiber materials based on the most important properties. It should be noted that these mechanical properties may vary considerably due to the large number of variants. This table only provides an initial overview.

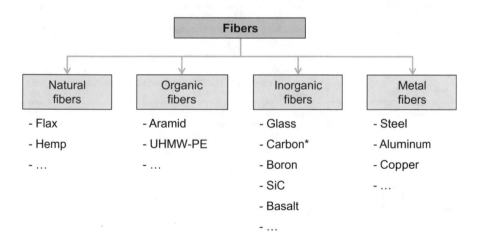

Fig. 4.21 Overview of relevant fiber materials for FRPs (the final carbon fiber represents a modification of almost pure carbon and is therefore categorized as inorganic; as the purity of the carbon fibers is not ideal, they are sometimes categorized as organic fibers)

Table 4.4 Profile "Glass fibers"

Glass fibers (GF)	
Description	Fibers with an amorphous molecular structure produced from molten glass by the melt-drawing process. Due to its favorable price-performance ratio (for E-glass) GF is by far the most important fiber material in FRP production
Important variants	• E-glass: standard aluminum borosilicate glass • S-glass: higher stiffness and strength compared to E-glass • C-glass: improved chemical resistance with slightly reduced elongation at break
To be considered	• High strength at low density • Relatively favorable price-performance ratio for E-glass • Significant price jump for S-glass • Comparatively high elongation at break allows design of flexible elements (e.g. leaf springs)
Typical fields of application in FRPs	Transport, building, rotor blades for wind energy plants, ship hulls, coiled and leaf springs, (chemical) tanks, molds, profiles

Table 4.5 Profile "carbon fibers"

Carbon fibers (CF)	
Description	Production from a precursor containing carbon (usually polyacrylnitrile) by stretch graphitization. The final carbon content can be adjusted by varying, among other things, the final temperature. Increasing graphitization leads to an increasing modulus of elasticity but after passing through a maximum, it leads to a decreasing strength
Important variants	• High tenacity (HT): highest strength • Intermediate modulus (IM): slightly higher stiffness with slightly reduced strength • High modulus (HM): stiffness further increased, strength further decreased • Ultra-high modulus (UHM): highest stiffness, strength further decreased
To be considered	• High modulus of elasticity at very low density • Low elongation at break • Extremely low, fiber-parallel thermal expansion (high-dimensional stability under thermal load, problematic when bonding to other materials • Risk of contact corrosion in connection with metals
Typical fields of application in FRPs	Sports vehicles, aeronautics, space (e.g. satellites), pressure vessels, high-performance sports articles

Table 4.6 Profile "aramid fibers"

Aramid fibers (AF)	
Description	Aramid stands for aromatic polyamide. Accordingly, AF belongs to the synthetic fibers. They are obtained from a solution by wet spinning, as the melting temperature is higher than the decomposition temperature. Afterward, they are stretched to the molecular orientation
Important variants	• Undrawn versus drawn: drawing results in molecular orientation and improved stiffness and strength in fiber direction
To be considered	• Relatively high tensile strength but low compressive strength at low density • High impact resistance and relatively high elongation at break • High energy absorption capacity • High temperature resistance • Negative fiber-parallel thermal expansion • High water absorption and low UV resistance • Low electrical conductivity
Typical fields of application in FRPs	Components at risk of impact, e.g. on boats, aircrafts, sports equipment, ballistic protection (e.g. helmets) and electrically insulating components

Table 4.7 Profile "polyethylene fibers"

Ultra-high-molecular-weight polyethylene fibers	
Description	Polymer fibers based on polyethylene with ultra-high-molecular-weight. Production by solution spinning and drawing
Important variants	–
To be considered	• High tensile strength in relation to density • High elongation at break • Low operating temperature • Fiber-matrix adhesion critical
Typical fields of application in FRPs	Components at risk of impact in the low temperature range, ballistic protection, can be combined with other fibers to increase the impact strength of the composite

In addition to the pure fiber properties, the structure of the fiber reinforcement plays a major role for the properties of FRP components. This will be discussed in the following section.

4.4.2   Structure of the Fiber Reinforcement

In Sect. 4.2 the relevance of fiber orientation and length for the mechanical properties of FRPs was already discussed. Both parameters cumulate in the structure of the fiber reinforcement. Figure 4.22 shows a ranking (trends) of common

Table 4.8 Profile "natural fibers"

Plant natural fibers (NF)	
Description	Obtained from plants (usually from the bast)
Important variants	• Hemp • Flax • Sisal • Jute
To be considered	• Properties are strongly influenced by the manufacturing process • Variations in properties must be taken into account • Very low density • Excellent acoustic damping properties • High potential for sustainable FRP production • High moisture absorption • Are only available in finite length before processing (usually to a non-woven
Typical fields of application in FRPs	Semi-structural components, e.g. in highly-priced vehicles with high demands on acoustic damping: door panels, interior parts, etc.

Table 4.9 Profile "basalt fibers"

Basalt fibers	
Description	Are obtained by melt spinning from volcanic basalt rock and have an amorphous structure
Important variants	The composition depends on the deposit of the basalt rock and affects the properties
To be considered	• Mechanical properties slightly higher than with E-glass, but more expensive • Relatively high temperature applications possible • Fibers are completely inert • High chemical and corrosion resistance
Typical fields of application in FRPs	Similar to glass fiber, but due to the higher price mostly only in niche applications where the higher temperature resistance is important

Table 4.10 Profile "boron fibers"

Boron fibers	
Description	Complex production by chemical vapor deposition on tungsten wire
To be considered	• Very high tensile and compressive strength and stiffness • Expensive production
Typical fields of application in FRPs	High-performance sporting goods, highly rigid components for military aviation

Table 4.11 Profile "steel fibers"

Steel fibers (SF)	
Description	Production by wire drawing process
To be considered	• Relatively high density • High elongation at break (high energy absorption) • High electrical conductivity
Typical fields of application in FRPs	Application so far only in research, to achieve multi-functionality (e.g. crash behavior, electrical conductivity); mass application in fiber-reinforced elastomers (e.g. carcass in tires)

reinforcement types with regard to specific (density related) properties. While the increasing performance from short fibers to long fibers and eventually mats is mainly achieved by increasing the fiber length and content, the performance of textile reinforcement is additionally increased by the possibility of exact fiber orientation and even further increased fiber volume contents. The best properties can be achieved if the fibers are optimally oriented according to the actual load situation. With regard to the lightweight design potential, however, it must also be taken into account that short/long fiber reinforcement offer superior possibilities in terms of structural lightweight design, e.g. by stiffening ribs, which are much more difficult to realize with a textile reinforcement. Figure 4.22 also shows a trend with regard to component costs. Even if the actual economic efficiency depends on further factors such as quantity, exact component geometry, etc., increasing fiber length and orientation often go together with higher semi-finished product costs, greater preparation effort and partly less efficient processes. As a result, there is a trade-off between the technical and economic requirements.

4.4.3 Determination of Mechanical Properties for the Preliminary Design

To decide which specific combination of fiber material and structure of the fiber reinforcement is suitable to meet the requirements, the mechanical characteristics of the corresponding combinations need to be known. To bypass the effort of an experimental determination of the relevant properties of all potential combinations there are two possible options:

1. Derivation of the mechanical properties of FRPs from the properties of the individual material partners (fiber and matrix).
2. Usage of database values.

The implementation of these options is explained in the following, whereby continuous and short/long fiber-reinforced polymers are considered separately.

Table 4.12 Comparison of properties of different fiber materials, partially from [13, 19–27] and different manufacturers data sheets

Fiber material		Density in cm^3	Tensile			T_{max} in °C	CTE in 10^{-6}/K fiber-parallel	Costs (tendency)
			Strength in GPa	Modulus in GPa	Elongation at break in%			
Glass fiber (GF)	E-Glass	2.54	2.4	73	4.8	300	5.1	Low
	S-Glass	2.49	4.5	87	5.7	250	5.6	High
Carbon fiber (CF)	HT	1.74	3.4	230	1.6	400–	−0.5	High
	IM	1.74	2.3	294	1.8	600	−1.2	High
	HM	1.81	2.5	392	0.7		−1.1	Very high
	UHM	1.90	2.2	450	0.4		−1.5	Very high
Aramid fiber	Standard	1.44	3.6	83	3.6	250	−2.0	Intermediate
	HM	1.44	3.6	124	2.4		−2.0	Intermediate
UHMW-PE		0.91	3.3	172	4.0	100	−12.1	Intermediate
Natural fibers	Hemp	1.45	0.6	70	1.6	200	n/a[a]	Low
	Flax	1.48	0.75	30	2.0			Low
	Jute	1.46	0.55	55	2.0			Low
	Kenaf	1.40	0.60	37	1.6			Low
Basalt fiber		2.70	3.7	110	3.2	700	6–9	High
Boron fiber		2.60	3.6	440	1.0	1800	4.9	Very high
Steel fiber (heat-treated)[b]		7.8	0.7	185	30	1000	10–16	Intermediate
Steel fiber (drawn)[b]		7.8	2.3	185	3.7	1000	10–16	Intermediate

[a]For natural fibers, thermal expansion is less relevant than swelling due to moisture absorption
[b]Data is partially based on unpublished results of IVW GmbH

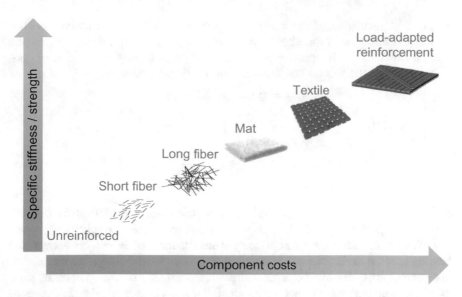

Fig. 4.22 Trends concerning the influence of the structure of the fiber reinforcement on the specific mechanical properties and the component costs

Continuous Reinforced Plastics

For the theoretical derivation of the mechanical properties, semi-empirical equations (mixing rules) are available, which allow derivation of the stiffness characteristics (the so-called engineering constants) of a continuous UD single layer directly from the material properties of fiber and polymer. For example [19] lists the following equations based on [28]:

- For the calculation of the stiffness $E_{\parallel}$ in fiber direction from the fiber longitudinal stiffness $E_{F\parallel}$, the matrix stiffness E_M and the fiber volume content φ:

$$E_{\parallel} = \varphi \cdot E_{F\parallel} + (1 - \varphi) \cdot E_M \tag{4.12}$$

- For the calculation of the stiffness transverse to the fiber direction:

$$E_{\perp} = \frac{E_M(1 + 0.85\varphi^2)}{\varphi \cdot \frac{E_M}{E_{F\perp}} + (1 - \varphi)^{1.25}} \tag{4.13}$$

- To calculate the shear stiffness $G_{\perp\parallel}$ from the matrix shear stiffness G_M, the fiber shear stiffness $G_{F\perp\parallel}$ and the fiber volume content φ:

$$G_{\perp\parallel} = \frac{G_M\left(1 + 0.6\varphi^{0.5}\right)}{\varphi \cdot \frac{G_M}{G_{F\perp\parallel}} + (1 - \varphi)^{1.25}} \tag{4.14}$$

- For the calculation of the transverse contractions v, from the transverse contraction behavior of the fiber v_F and the matrix v_M, the following applies (the first subscript of the transverse contraction figures gives the direction of contraction and the second one the direction of the applied load):

$$v_{\perp\|} = \varphi \cdot v_{F\perp\|} + (1 - \varphi) \cdot v_M \tag{4.15}$$

respectively

$$v_{\|\perp} = v_{\perp\|} \cdot \frac{E_\perp}{E_\|} \tag{4.16}$$

The data for the fiber and matrix can be taken from the literature or from appropriate databases. Table 4.12 provides a first overview of mechanical properties for different fiber materials. Since the matrix material is not yet known and bearing in mind the dominance of the fiber properties for the composite mechanics, the matrix properties can be neglected at this point. If the fiber properties were extracted from the manufacturer's data sheet, one must keep in mind that the tests to provide these values are carried out on ideal fibers and under ideal conditions, which cannot be achieved in industrial reality, e.g. due to imperfections concerning fiber parallelism in the fiber bundle. Furthermore, when applying these formulae, the fiber volume fractions should not be assumed too high, so that they can actually be achieved in the real part. For example, not more than 55% fiber volume content should be assumed for textiles and not more than 65% fiber volume content for pure UD structures.

To minimize the effort, it is advisable to use programs in which these or similar formulas are already implemented. For example, the alfaLam[3] program for laminate analysis, provided by the Technical University of Darmstadt (Department of Lightweight Structures and Conceptual Design) is free of charge. It is based on the formulas given in VDI 2014 [8].

With the procedure shown, the stiffness behavior of a unidirectional, continuously reinforced structure can be derived. However, since very few components are purely unidirectional reinforced, the question arises how the mechanical behavior of the overall laminate can be determined. The answer is provided by the **classical laminate theory** (CLT). With the CLT, the mechanical behavior of a multilayer laminate can be derived from the mechanical characteristics of the single laminate layers and their orientation to each other. It thus serves to determine the law of elasticity. On this basis, the distortions and stresses of the laminate under load can be calculated. The CLT follows the procedure shown in Fig. 4.23.

---

[3]http://www.klub.tu-darmstadt.de/forschung_klub/downloads_3/downloads_klub.de.jsp, downloaded on 06.08.2018.

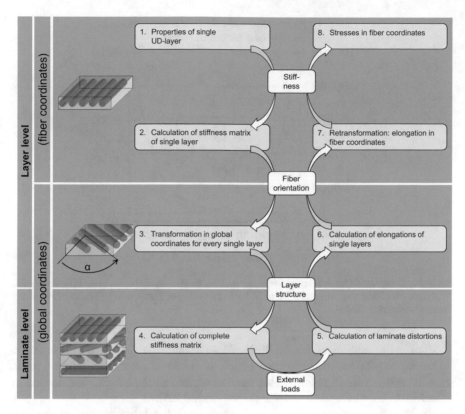

Fig. 4.23 Basic procedure of the classical laminate theory (CLT). Adapted from [19] (Image adapted, printed with permission of Springer Nature)

This procedure is the basis of several available analytical and FE-based programs. In detail, the procedure that can be seen in the figure is the following [19]:

1. The starting point is given by the engineering constants, which describe the behavior of the single UD layer under uniaxial loading or pure shear. They can be calculated by Eqs. (4.12)–(4.16).
2. From the engineering constants of the single UD-layers the stiffness matrix (3 × 3) can be derived, which describes the material behavior under multiaxial loading. This can be done for all layers within a laminate. Figure 4.24 shows an example of the directionally transformed engineering constants of a single UD layer of a carbon fiber-reinforced polyamide 66.
3. This stiffness matrix is then converted from the fiber coordinate system to the laminate coordinate system. This is done by forming the corresponding transformation matrix for each layer.

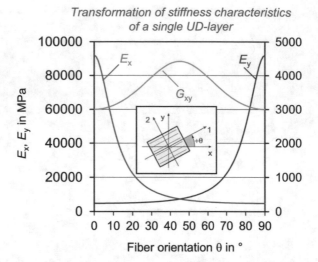

Fig. 4.24 Direction-dependent properties of a UD single layer of carbon fiber-reinforced polyamide 66 with a fiber volume content of 55%. Printed with permission of Leibniz-Institut für Verbundwerkstoffe GmbH

4. Considering additional data on the laminate structure (sequence and thickness of the layers) the laminate stiffness matrix is obtained, often also referred to as ABD matrix (see Fig. 4.25). This matrix contains all the information needed to calculate the mechanical couplings (see Sect. 4.2.3).
5. By inversion, the stiffness matrix becomes the flexibility matrix. Using this matrix, the distortions resulting from external loads can be calculated.
6. Assuming that the individual layers are firmly bonded to each other, the distortions of the laminate can be used to derive the strains of the individual layers.
7. These strains can then again be converted from the laminate coordinates into the fiber coordinates.
8. From the strains of the single layers, using the stiffness matrix of the single layers, it is possible to derive the stresses in the single layers.

In addition to this approach, for simple calculations during concept development, the possibility of deriving the engineering constants of the laminate from the ABD matrix is of interest. These engineering constants then describe the stiffness behavior in a certain direction, under uniaxial load. Figure 4.26 shows as an example, engineering constants for different laminates of carbon fiber-reinforced polyamide 66, as a function of the load angle in relation to the main fiber angle. The laminate [0/+45/90/−45]$_s$, a classic quasi-isotropic laminate, shows almost no influence of the load angle.

Concerning the detailed mathematical description of the procedure, the reader is referred to the vast amount of existing literature (e.g. [13]). A basic understanding

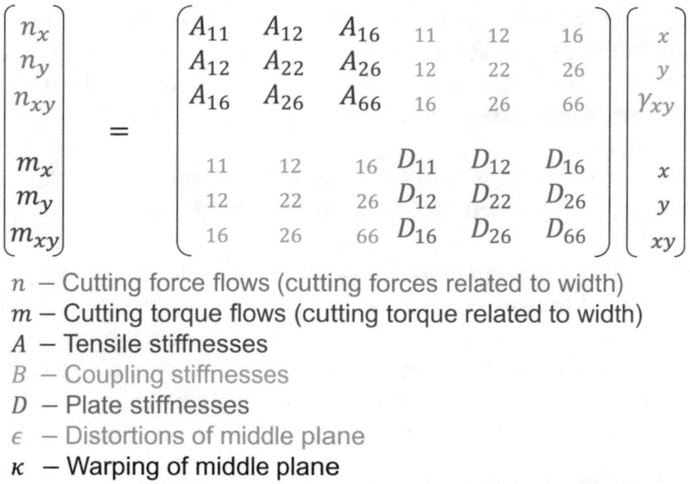

$$\begin{pmatrix} n_x \\ n_y \\ n_{xy} \\ m_x \\ m_y \\ m_{xy} \end{pmatrix} = \begin{pmatrix} A_{11} & A_{12} & A_{16} & {\scriptstyle 11} & {\scriptstyle 12} & {\scriptstyle 16} \\ A_{12} & A_{22} & A_{26} & {\scriptstyle 12} & {\scriptstyle 22} & {\scriptstyle 26} \\ A_{16} & A_{26} & A_{66} & {\scriptstyle 16} & {\scriptstyle 26} & {\scriptstyle 66} \\ {\scriptstyle 11} & {\scriptstyle 12} & {\scriptstyle 16} & D_{11} & D_{12} & D_{16} \\ {\scriptstyle 12} & {\scriptstyle 22} & {\scriptstyle 26} & D_{12} & D_{22} & D_{26} \\ {\scriptstyle 16} & {\scriptstyle 26} & {\scriptstyle 66} & D_{16} & D_{26} & D_{66} \end{pmatrix} \begin{pmatrix} x \\ y \\ \gamma_{xy} \\ x \\ y \\ xy \end{pmatrix}$$

n — Cutting force flows (cutting forces related to width)
m — Cutting torque flows (cutting torque related to width)
A — Tensile stiffnesses
B — Coupling stiffnesses
D — Plate stiffnesses
ϵ — Distortions of middle plane
κ — Warping of middle plane

Fig. 4.25 Build-up of a laminate stiffness matrix (often referred to as ABD matrix)

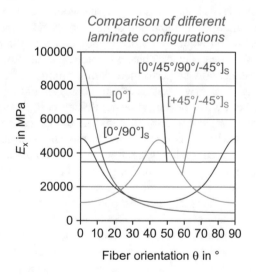

Fig. 4.26 Direction-dependent properties of several laminate configurations of a carbon fiber-reinforced polyamide 66 with a fiber volume content of 55%. Image adapted, printed with permission of Leibniz-Institut für Verbundwerkstoffe GmbH

of the fundamentals of the CLT is indispensable for the correct design of FRP components. However, from an application-oriented perspective, the main interest is on how the CLT can be applied as efficiently as possible. Since the complexity of the calculation increases rapidly with an increasing number of layers, the use of software is appropriate to accelerate this step. A very good starting point is given, for example, by another free program eLamX^{24} developed by the Chair of Aircraft Engineering at the Technical University of Dresden. It is based on the CLT and enables, among other things, the fast strength and stiffness analysis of laminates including the engineering constants (Fig. 4.27). Couplings (see Sect. 4.2.3) can also be considered [29].

With these methods, it is possible to calculate quite quickly, the mechanical properties of laminates based on the mechanical properties of the fibers and the matrix material.

A second option, allowing one to bypass the effort of a calculation, could be to take values from databases. Various corresponding databases are available (some free-of-charge, some fee-based), for example www.campusplastics.com, Total Materia, Granta Design or the integrated database in the elamX^2 program. If this strategy is followed, common material systems should be chosen. In addition, the values should be critically questioned, for example through cross-comparisons.

Short or Long Fiber-Reinforced Polymers

In the case of short or long fiber-reinforced FRPs, the application of the rules of mixtures given in Eqs. (4.12)–(4.16), leads to an overestimation of the values achievable in reality. Hence, the rules of mixture would have to be extended by a reduction factor, resulting from a function of fiber orientation, fiber-matrix adhesion and the fiber aspect ratio (length to diameter) [30]. Defining the function would be accompanied by large uncertainties. Thus, a theoretical estimation is not appropriate for the initial calculations to be performed during product conception. Therefore, database values (e.g. from www.campusplastics.com) should be used for a simple estimation of whether a short or long fiber-reinforced design could be sufficient to meet the requirements regarding critical loads. As the matrix polymer has not yet been determined and will not be determined at this stage, common standard materials should be considered, such as glass fiber-reinforced polyamide 66. Due to the uncertainty of this approach, short or long fiber reinforcements should only be chosen if the requirements are very clearly met by the database values, because in the real-life component, the mechanical properties can be significantly lower. In the case of values from widely available data sheets, it must be noted, that, e.g. tests for injection molding compounds are typically carried out under idealized conditions, e.g. on a tensile bar which has a comparatively high anisotropy. The resulting values are therefore not achieved in most areas of the later products. In addition, there are strong influences, for example, through moisture absorption.

With the mechanical properties generated this way, the selection process for fiber materials and the structure of the fiber reinforcement can now begin. Over time, an own database of mechanical properties is built up which can be used repeatedly, so that this step can be carried out with increasing efficiency.

---

[4]Web link: https://tu-dresden.de/ing/maschinenwesen/ilr/lft/elamx2/elamx.

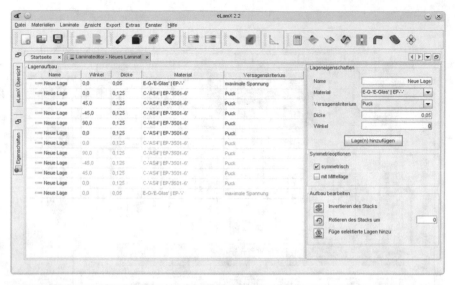

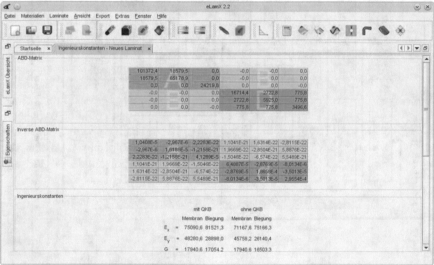

Fig. 4.27 Software "eLamX²"—Interface for the definition of a laminate (top) and output of the corresponding ABD-matrix and engineering constants (bottom). Images adapted, printed with permission of TU Dresden, Institut for Aerospace Engineering

4.4.4   Selection Procedure

In the following section, it is shown, how a decision regarding the combination of fiber material and structure of the fiber reinforcement can be made. In this context, cost-efficiency is the highest priority, which means that the cheapest solution meeting all the requirements is sought. Figure 4.28 shows the proposed procedure.

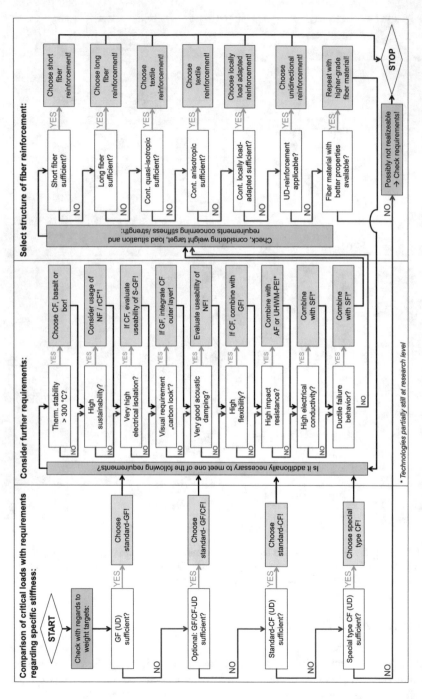

Fig. 4.28 Selection procedure for fiber material and structure of the fiber reinforcement

In a first step, it is checked, which fiber material is best suited to achieve the required weight and stiffness targets, with regard to the identified critical loads (see Sect. 4.3). Initially, for each load case—a pure UD design is assumed, in which the fibers are ideally oriented in relation to the load case. The mechanical properties for this evaluation are determined according to the methods introduced in Sect. 4.4.3.

Figure 4.28 shows a clear preference order for glass fiber (E-glass) and various types of carbon fiber. Since GFRPs and CFRPs dominate by far the market, these materials should be seen as standard materials. As such, they go together with a corresponding risk minimization and are therefore prioritized. Independent of the structure of the fiber reinforcement, GF is preferred over CF. Due to the large price differences, a GFRP component will almost always be cheaper than a CFRP component. UD designs are considered, because with regard to the individual load cases they represent the best case, making full use of the fiber properties. Hence, if the requirements cannot be achieved with a specific fiber material in a UD design, there is no other way than to choose a higher-performance fiber.

Once the fiber material has been selected in this way, the second step is to check whether there are any further requirements that require an alternative fiber selection. This could lead to complete replacement of the first selected fiber material or to a combination of fiber materials. For example, to increase damage tolerance against impacts, the integration of aramid fibers (AF) could be an option. The presented list of possible requirements does not claim to be complete, as various requirements are possible, such as high radiation transparency. At this point, the team member from the materials science department should therefore be involved in order to keep an eye on all requirements and to make use of his expertise in the selection of materials.

Once the fiber selection has been completed for the time being, it is checked which structure of the fiber reinforcement is suitable to meet all stiffness and strength criteria. In accordance with the explanations in Sect. 4.4.2, a clear prioritization is made here, in terms of the cost-efficiency. In order to be able to check whether the requirements are met by a specific structure of the fiber reinforcement, mechanical properties are required, which are determined as explained in Sect. 4.4.3. If even a highly anisotropic design based on textile reinforcements is not sufficient to meet the requirements, a locally load-adapted structure with varying wall thickness might be an option, but this should then be discussed with the team member from the manufacturing department. If in fact only the unidirectional design is suitable to meet the requirements, it should be noted that this is rarely actually feasible, since it presupposes that apart from the critical load there are no other relevant loads that act at a different angle.[5] Therefore, if no structure of the fiber reinforcement is mechanically sufficient, apart from the pure UD design, and if this is not practical, for example due to relevant transverse forces, the selection process is repeated with a higher-performance fiber material (according to the order of priority shown in Fig. 4.28). If, for example, initially a widely used

---

[5]Leaf springs are a good example of an application that meets this requirement [31].

standard carbon fiber was selected, the next higher-performing fiber material would be given by a less common special form (e.g. ultra-high modulus, see Table 4.5). It can happen that not all requests can be met the same time. In this case, they must be critically questioned again.

The described procedure is now illustrated using the example of the thread lever. In a first step, the critical loads are used to check, whether a GF design with unidirectional reinforcement allows achieving the weight targets. For this, Table 4.13 shows exemplary values for different fiber materials and structures of the fiber reinforcement that can be used for a first estimation. In order to determine the suitability of the various fiber materials, these mechanical properties are used to derive the cross-sectional areas required to achieve the stiffness targets (see Sect. 4.3). Figure 4.29 shows examples for the procedure.

In accordance with the method shown in Fig. 4.28, the first step is to derive whether the weight targets can be achieved with the UD-GFRP. For this purpose, the resulting weight is to be calculated. Of course, many designers would already use simple CAD models for this purpose, but often a simple manual calculation can be sufficient at this stage. Figure 4.18 shows the fixed geometries, consisting of the two bearings, the thread eyelet and the relative positions to each other. The component must connect and frame these three components, whereby the height of the

Table 4.13 Example values for different fiber materials and structures of fiber reinforcement

Structure of fiber reinforcement	Fiber material	Fiber volume content (%)	Fiber orientation	Density[a] (g/cm^3)	E_x (MPa)	E_y[b] (MPa)
UD-reinforced polymer	GF	55	UD	1.89	40,150	–
	CF (HT)	55		1.45	126,500	–
Textile-reinforced polymer[c]	GF			1.82	18,113	18,113
	GF	50	[0/+45/90/ −45]$_S$	1.82	13,550	13,550
·	CF (HT)	50	[0/90]$_{2S}$	1.42	56,925	56,925
	CF (HT)	50	[0/+45/90/ −45]$_S$	1.42	38,813	38,813
Short fiber-reinforced polymer[d]	GF	14	Quasi-isotropic	1.14	6,700	6,700
	CF (HT)	14		1.22	11,000	11,000

[a]The mixing density results from the density of the fiber material and the matrix material under consideration of the volume fractions, a density of 1.1 g/cm^3 is assumed for the matrix
[b]For UD materials, the lateral stiffness is not calculated because the matrix is neglected and the result would therefore be zero
[c]Reduction factors were deducted, lateral stiffness was neglected, thus creating an additional safety buffer
[d]Database values for short fiber-reinforced PP (GF) and PA66 (CF)

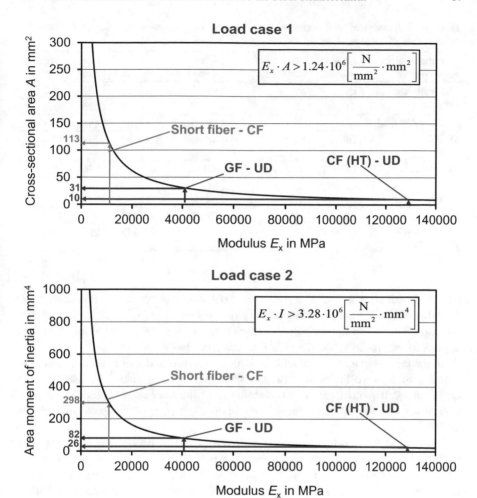

Fig. 4.29 Derivation of the required geometry for different FRPs

connection between bearing and thread eyelet is also limited. From the curves shown in Fig. 4.29, the cross section of the GFRP between the bearings and the required area moment of inertia between the bearing and the thread eyelet can be derived. Based on the distance between the bearings and the cross section required there, the volume required for the connection can be calculated. The volume between the bearing and the thread eyelet can be determined by the distance, the required moment of inertia and the specification of a maximum height of 4.5 mm (the maximum height is fully utilized, to maximize the bending stiffness). To account for the framing of the bearings and the eyelet, the volume is increased by 20%. For the UD-GFRP design, the result is a weight of 4.3 g. The aim of the development was to reduce the weight compared to the component weight of 5.2 g,

Table 4.14 Estimation of the part weight that can be reached with different FRPs under consideration of the critical tensile and bending load, respectively

Structure of fiber reinforcement	Fiber material	Fiber volume content (%)	Fiber orientation	A in mm^2	l in mm^4	Weight in g
UD-reinforced polymer	GF	55	UD	31	82	4.3
	CF (HT)	55		10	26	1.0
Textile-reinforced polymer	GF	50	$[0/90]_{2S}$	68	181	9.2
	GF	50	$[0/+45/90/-45]_S$	92	242	12.2
	CF (HT)	50	$[0/90]_{2S}$	22	58	2.3
	CF (HT)	50	$[0/+45/90/-45]_S$	32	85	3.3
Short fiber-reinforced polymer	GF	14	Quasi-isotropic	185	490	15.5
	CF (HT)	14		113	298	10.1

achieved with aluminum die-casting. This goal would therefore be achievable with the UD-GFRP. The next step foreseen by the procedure in Fig. 4.28 is to check the selection concerning further possible requirements. None of the listed requirements applies to the thread lever; the GFRP is therefore basically suitable.

In the third step it is now examined which structure of the fiber reinforcement is required to achieve the weight targets with the GFRP. Table 4.14 lists various GFRP materials, which differ in the structure of the fiber reinforcement. The procedure for the weight calculation is equal to the one applied for unidirectional GFRP.

Looking at the short glass fiber-reinforced polymer, it becomes clear that the estimated component weight of around 16 g is well above the reference of 5.2 g. Even for an alternative material, that for example has a stiffer matrix or a higher fiber volume content, the achievement of the target is unlikely, so that this option can be excluded. As further listed in Table 4.14, a GFRP with textile reinforcement is also not promising. Only the UD design is below the reference weight. Since up to now only the critical load cases have been considered, it must now be questioned, whether a UD design can actually be realized. Since the analysis of the thread lever also revealed considerable forces transverse to the direction of the critical load, a pure UD design cannot be realized. Given the small difference between the weight of the component in GFRP-UD design and in aluminum die-cast design, and because the component dimensions are relatively small, it cannot be expected that a design with locally load-adapted reinforcement will lead to the desired result. Therefore, according to the order of priority shown in Fig. 4.28, a higher-performing fiber material is chosen. As the weight savings, achieved with the UD-GFRP are rather small, the possibility of a GFRP-CFRP hybrid is not considered, and the potential of CFRPs is directly examined. With CF as the new base material, all further requirements should again be examined. In the case of the thread lever, there is no requirement that speaks against application of CF. The weights that can be reached with different structures of the fiber reinforcement are again listed in Table 4.14. The short fiber-reinforced variant is far above the reference value, but for all continuously reinforced variants, there is a clear potential of weight savings. According to the given order of priority, a quasi-isotropic textile reinforcement could therefore be selected. With a weight of 3.3 g, the weight-related targets are likely to be achieved, even when all other occurring forces are considered.

The presented selection methodology can of course only show tendencies. Because of the many interdependencies between the material, the process, the component geometry and the economic conditions, no generally applicable methodology can be established here. The intention behind the presented method was therefore to make clear that the principles of integrated product development are especially relevant for the selection of the fiber material and the structure of the fiber reinforcement, in order to ensure an optimal product design.

4.5 Pre-dimensioning

Based on the principal selections made in the previous steps, a first FE model can be built-up, with which the selection can be validated and the geometry can be roughly pre-defined. This minimizes the risks in the subsequent development of the process concept. Figure 4.30 shows an FE model including a deformation analysis for the two critical load cases.

The deformation analysis shows that the quasi-isotropic design is preferable, with regard to the load cases. Accordingly, the material selection can be refined.

4.6 Development of a Process Concept

The pre-dimensioning, the decisions on the fiber material and the structure of the fiber reinforcement, and the requirements catalog set the basis for the development of a process concept for the manufacturing of the component. According to Fig. 4.1, the team member from the area of manufacturing technology is responsible for this action.

In the following, at first, the fundamentals of FRP manufacturing are presented. Subsequently, the most important manufacturing processes are described, and finally, a procedure for the selection of processes is presented.

Finite element model of the CFRP - thread lever:

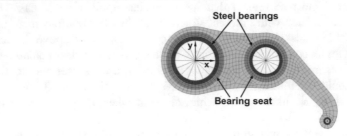

Results of the deformation simulation:

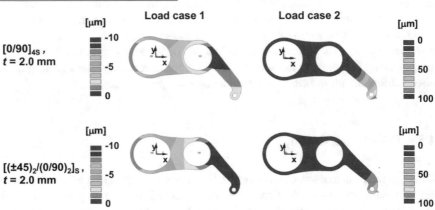

Fig. 4.30 Finite element model of the CFRP thread lever and deformation analysis for the two critical load cases. Images adapted, printed with permission of Leibniz-Institut für Verbundwerkstoffe GmbH

4.6.1 Fundamentals of FRP Manufacturing

Producing FRPs means that the fiber structure and the matrix must be combined together. Regardless of the semi-finished product stages into which the process is divided or which specific manufacturing method is used, there are three decisive process steps [20]:

- **Impregnation** describes the process of wetting the individual filaments and filling the spaces between the filaments with liquid matrix.
- During **consolidation**, the target fiber volume content is set by pressure application, which also counteracts restoring forces. If air is present in the reinforcement structure, it is driven out and the formation of new air inclusions is prevented. If the FRP is a laminate, the individual layers are bonded to each other.
- **Solidification** describes the transition of the liquid matrix into the solid state, either by solidification (thermoplastics) or chemical cross-linking (thermosets). In contrast to thermosets, the solidification of thermoplastics is reversible; i.e. they can be melted again (see Sect. 4.7).

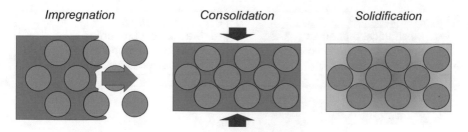

Fig. 4.31 Illustration of the three core processes of FRP manufacturing

Figure 4.31 illustrates these steps. Consolidation and solidification often take place simultaneously, and, depending on the process, this is also true for consolidation and impregnation.

According to Fig. 4.32 it is always the process variables

- **temperature**,
- **pressure** and
- **time**

that determine the course of the process and the process result (mechanical and morphological properties of the FRP), due to their influence on the processing behavior of the fiber structure (see Sect. 5.3.1.4) and the matrix (see Sect. 5.3.2.2) [20].

In the following section, the most important manufacturing processes are presented. In order to understand the explanations of the manufacturing processes, it is important to consider the fundamental **difference between thermoset- and thermoplastic**-based processes. Thermosets polymerize directly in the process, starting from monomers or short-chain oligomers. As a result, their viscosity is initially very low, and therefore, quite large flow distances within the reinforcement structures are possible. Yet, the polymerization and solidification are irreversible. Thermoplastics on the other hand can be re-melted; i.e., casting and forming processes can be separated. However, melted thermoplastics consist of long polymer chains. The viscosity is therefore many times higher than that of thermosets, and the flow paths that can be covered within a reinforcing structure are relatively small. These different conditions often lead to fundamentally different approaches in FRP manufacturing.

Another important term to be aware of is "prepreg" which is derived from the English term "pre-impregnated" Prepregs are semi-finished products for the production of FRPs, in which fibers and matrix material have already been combined. Prepregs are available in a wide variety of designs, as presented in detail in Sect. 5.3.1.2.

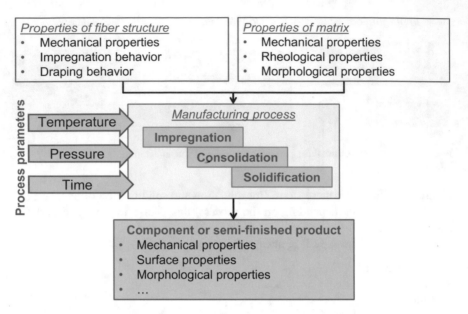

Fig. 4.32 Fundamentals of FRP manufacturing. Adapted from [20] (Printed with permission of Carl Hanser Verlag GmbH & Co. KG)

4.6.2 Manufacturing Processes

Resulting from boundary conditions, such as the component geometry, the structure of the fiber reinforcement and the flow properties of the matrix polymers, a large number of processing chains for FRPs have been established. They can be separated into direct and indirect processing chains. Direct processing chains are characterized by the fact that dry semi-finished fiber products (e.g. textiles) are combined with the matrix in a single process, in which the component is directly manufactured. Indirect process chains are based on semi-finished fiber/matrix products, so-called prepregs, in which the reinforcing structure and the matrix are already combined. Figure 4.33 shows the most important processing chains for the production of FRP components. The starting point is always given by the fibers, which are initially converted to 1D semi-finished products (mainly rovings, i.e. fiber bundles). In the illustration, dry semi-finished products are highlighted in gray, prepregs in light blue and the actual manufacturing processes in dark blue.

The figure intends to show the most important production chains. Besides these, there are of course other possibilities to produce FRP components, which were not included, for the sake of clarity. This applies, e.g. to the various process combinations, which will be discussed in more detail in Sect. 4.6.2.15. In the following section, the individual production chains are briefly introduced, in order to establish the basic knowledge required for process selection, the selection of semi-finished products and eventually an appropriate design to manufacture.

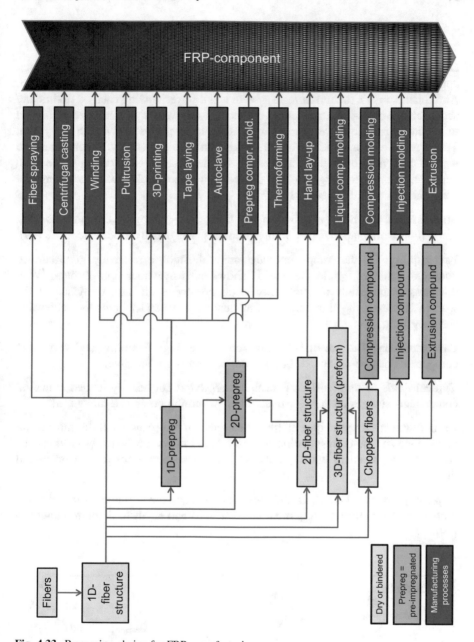

Fig. 4.33 Processing chains for FRP manufacturing

4.6.2.1 Fiber Spraying

Description: In fiber spraying, rovings are fed to a spray gun (Fig. 4.34), which contains a cutting unit that shortens the rovings to a length of typically 12–25 mm. At the same time, a thermoset resin system and a corresponding catalyst (to initiate/accelerate the cross-linking reaction) are fed into the system. Compressed air is used to the cut fibers and drive them as well as the resin/catalyst out of the spray gun. This way the resin and the catalyst are mixed, and simultaneously, the cut fibers are wetted. Spraying onto the tool results in the desired component shape and, depending on the spraying time, the desired component thickness. After curing, the component can be demolded [19, 32].

Structure of the fiber reinforcement: Long fiber-reinforced components with random fiber orientation. Fiber volume contents are typically around 15–20% [19].

Part quality: On the component side facing the tool, a relatively good surface quality is achieved, while the outside shows extensive fiber print-through. Wall thickness homogeneity and tolerance strongly depend on the experience of the worker. However, the high amount of manual work generally limits the achievable accuracy [33].

Part geometry: The use of the spray gun ensures high flexibility and very large components can be produced. In addition, undercuts are possible.

Typical fields of application: Especially larger but not heavily loaded structural components such as containers, boat hulls, swimming pools and tanks [33].

Input materials: In principle, all fiber materials can be processed, if the cutting unit is designed for the corresponding material. Glass fibers are typically used, so the corresponding standard cutting units are often not suitable for carbon or aramid fibers.

Typical quantities: Due to its high flexibility, the process is suitable for individual production (one-off parts, e.g. mold construction) and small to medium quantities [19].

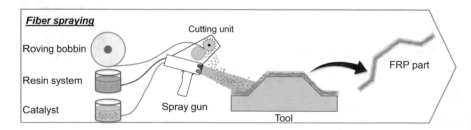

Fig. 4.34 Schematic illustration of the process "fiber spraying"

To be considered: The method allows a layered design with covering surface layers of pure resin and supporting laminates. Due to the high amount of manual work, experienced personnel is required, in order to reach reproducible quality [19].

4.6.2.2 Centrifugal Casting

Description: A cylinder is set in rotation along the longitudinal axis (Fig. 4.35). Inside the cylinder, rovings and a thermoset resin system are fed into a spray nozzle. In the lance, the rovings are shortened using a cutting unit. The fibers are then driven out via compressed air, together with the resin system. The mixture is sprayed on the inside of the rotating cylinder and distributed evenly by the centrifugal force. Optionally, textiles can also be added [34, 35].

Structure of the fiber reinforcement: Short fiber-reinforced with random orientation or with integrated textile semi-finished products, such as mats or even fabrics. The fiber reinforcement is compressed by the centrifugal force, resulting in a fiber volume content of about 30% with chopped fibers and about 40% when fabrics are used [19].

Part quality: Since the centrifugal force leads to a separation of materials according to density, a mirror-smooth pure resin layer is created on the inside, which protects the laminate [36].

Part geometry: Depending on the type of reinforcement, the process can be used for pipes with a diameter ranging from a few centimeters to several meters, whereby the rotation speed required to achieve the centrifugal forces needed for good component quality sets technical limits. If the pipes become too small, the feeding of the materials becomes difficult. The pipes can be several meters long [35, 37].

Typical fields of application: Pipes for the building industry, tanks, pressure pipelines, but also conical masts [19].

Input materials: Textiles or rovings (almost exclusively glass fibers) and resins (typically UP, VE and EP).

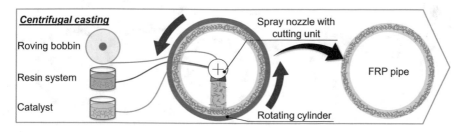

Fig. 4.35 Schematic illustration of the process "centrifugal casting"

Typical quantities: Due to the necessary procurement of the cylindrical mold on the one hand and the required cycle times for full curing on the other hand, centrifugal casting is especially suitable for small to medium quantities [37].

To be considered: As a process for rotationally symmetrical components, this process is partly in competition with winding and pultrusion. Therefore, all three manufacturing processes must always be considered with regard to their suitability.

4.6.2.3   Winding

Description: A rotating core pulls rovings from a spool stand. A CNC-controlled, position-variable ring thread eye creates a winding pattern according to the mechanical requirements. In classical wet winding, the roving is pulled through an impregnation unit, using the pull-off forces generated by the tool rotation. In the impregnation unit the roving is impregnated with a thermoset resin system (see Fig. 4.36). The component is then usually subjected to a subsequent curing step in an oven. Alternatively, thermoset or thermoplastic pre-impregnated rovings (so-called tapes) can be used, whereby possibly a preheating station and an additional consolidation roller are required for the thermoplastic variant. Finally, also dry winding and subsequent liquid impregnation with a thermoset resin is possible. However, it should be noted that the impregnation behavior of the wound structure is quite challenging (see Sect. 5.3.1.4) [38–41].

Structure of the fiber reinforcement: Continuous fiber-reinforced, whereby the required geodetic paths lead to restrictions. For example, fiber orientation parallel to the axis of rotation (0° for cylinders) is only possible if pins are attached to the cylinder ends, around which the roving can be wound. Due to the high yarn tension, fiber volume contents of 40–70% can be achieved [20].

Part quality: Due to the process, the fiber orientation is relatively accurate and the use of rovings and the high roving tension result in relatively high fiber volume contents. During wet winding, a protective pure resin layer is automatically formed on the outside.

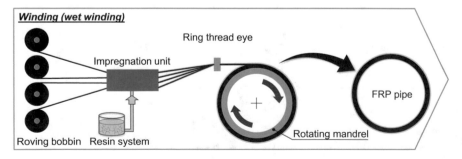

Fig. 4.36 Schematic illustration of the process "winding" (process variant wet winding)

Part geometry: There is a far-reaching restriction to rotationally symmetrical bodies (e.g. cylinders, cylinders with dome caps, ellipsoids, cones), whereby the winding of plates is also possible. In addition, there are special technologies that allow, for example, the winding of pipe branches or the winding of cylinders with integrated stiffening truss structures (so-called isogrids). Typical diameters for pipes and tanks range from 1 to 5 m, but larger components are also possible [20, 42–44].

Typical fields of application: Pressure vessels, drive shafts, rollers, e.g. for paper industry, insulators, crash elements, struts.

Input materials: Rovings (all fiber materials) in combination with thermoset resin systems or pre-impregnated tapes (thermoset and thermoplastic).

Typical quantities: Due to the required liner or winding core, the process is unsuitable for one-off production. Typical batch sizes range from small series up to automated large series production with an output of up to 300,000 parts per year. Comparatively high production efficiency can be achieved by using multiple winding systems, which can comprise several winding cores, so that several components can be wound simultaneously [19].

To be considered: As a method for rotationally symmetric components, this method is partly in competition with the centrifugal casting and pultrusion. Hence, for corresponding parts all three manufacturing processes must be considered, with regard to their suitability.

4.6.2.4 Extrusion

Description: During extrusion, a granulate of cut, unidirectional fiber-reinforced thermoplastic is melted and plasticized in an extruder (Fig. 4.37). The screw used for this purpose simultaneously pushes the material through a forming unit in which a defined cross section is generated. The thermoplastic solidifies by cooling below melting temperature. The continuous production thus allows the manufacture of FRP profiles of different lengths [30].

Structure of the fiber reinforcement: Short fiber-reinforced, whereas partially a flow-induced preferential orientation may occur [45]. The fiber volume content typically ranges from 25 to 30%.

Part quality: The extrusion process usually results in a rather rough surface [46].

Part geometry: Limited to straight profiles, but complex profile cross-sections are possible due to the short fiber reinforcement.

Typical fields of application: Extrusion is by far the leading process for the production of components made of wood-plastic composites, so-called wood polymer composites (WPCs). The majority of these components are used in the building industry, especially as terrace floors [47].

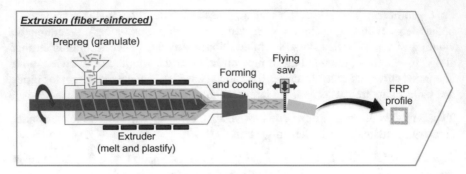

Fig. 4.37 Schematic illustration of the process "extrusion"

Input materials: Granules of unidirectional fiber-reinforced thermoplastic strands.

Typical quantities: As a continuous process, which is widely used in the pure plastics industry, the extrusion process is suitable for mass production. If multiple products are produced on a single production system, in order to operate it close to the capacity limit, even medium quantities can be produced economically, since only a tool must be procured and exchanged for each specific product.

To be considered: The process allows the production of uniaxial profiles and is therefore partly in competition with pultrusion.

4.6.2.5   Pultrusion

Description: The term pultrusion is derived from the English words "pull" and "extrusion." Accordingly, rovings are pulled from a spool and subsequently through an impregnation unit (thermoset) and a forming and curing unit (Fig. 4.38). The pulling force is applied by a pulling unit (e.g. caterpillar puller), which acts on the already cured FRP. This way, different cross-sections with continuous, unidirectional fiber reinforcement can be continuously pultruded. Depending on the cross-section, the feeding of textile semi-finished fiber products is also possible, but only close to or on the outer contour. In addition to the classic thermoset variant, there is a thermoplastic alternative in which hybrid rovings ("commingled yarns"), made of reinforcing fibers and thermoplastic fibers, are drawn through a melting unit and a forming and cooling unit [48, 49].

Structure of the fiber reinforcement: Typically, endless unidirectional fiber-reinforced. Reinforcements perpendicular to the pulling direction are partly possible by inserting textile reinforcements [50] (only close to or on the outer contour). Newer process variants such as pull winding or pull braiding aim at an even stronger integration of cross-reinforcing fibers by adding a winding or braiding process to the pultrusion process [51]. With unidirectional reinforcement, very high fiber volume contents of up to 80% can be achieved [52].

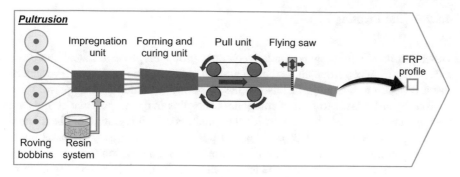

Fig. 4.38 Schematic illustration of the process "pultrusion" (thermoset)

Part quality: Due to the high fiber volume contents, the surface quality is usually relatively low [53], but can be improved by integration of nonwovens on the outer contour.

Part geometry: Largely limited to single-axis profiles (no change in cross-section). Through the use of so-called flying cores, it is also possible to manufacture hollow profiles [20]. Newer concepts also allow the production of multi-curved profiles through a movable forming unit ("moving mold process") [50, 54, 55].

Typical fields of application: Pultruded GFRP profiles are, e.g. used in the transport, the storage and the building sector (e.g. window/door frames and thresholds) and due to the combination of weather resistance, electrical insulation and electromagnetic permeability they can also be found in electrical engineering applications (e.g. insulator bars for overhead lines and mounting profiles for transformers and radomes). Pultruded CFRP profiles are also used for sports equipment (e.g. ski poles, fishing rods) [50], as belts for blades of wind turbines [19] and as floor cross-beams used for aircrafts [56].

Input materials: Rovings in combination with a thermoset resin system, less frequently "commingled yarns" made of reinforcement and thermoplastic fibers. The use of the thermoset alternative offers very limited possibilities for processing of textile semi-finished products.

Typical quantities: As a continuous process, the pultrusion process is suitable for mass production. If multiple products are produced on a single production system, in order to operate it close to the capacity limit, even medium quantities can be produced economically, since only a tool must be procured and exchanged for each specific product.

To be considered: As a process that allows the manufacture of rotationally symmetric components, this process is partly in competition with winding and centrifugal casting. Therefore, for corresponding components, all three manufacturing processes must always be evaluated, whereas special attention must be paid to the required structure of the fiber reinforcement.

4.6.2.6   3D Printing

Description: 3D printing is a relatively new group of processes, and there are indeed some concepts that allow the integration of fibers. One of them is a variant of fused deposition modeling. In this process, a plastic is melted (thermoplastic) and stacked layer by layer to form a component. For this purpose, polymer strands are conveyed into a heated nozzle in which they melt. For the integration of fibers, there are two possibilities (Fig. 4.39). Either the polymer strand already contains fibers (short or continuous fibers) or there is an inline melt impregnation (only continuous filaments). Both processes offer the technically very interesting possibility to form the component freely within the design space; i.e. without the need for a tool, it is possible to produce, for example, grid structures [57, 58].

Structure of the fiber reinforcement: Short or continuous fibers. The fiber volume content is typically around 30% [58].

Part quality: Currently relatively low (high porosity, air inclusions), as the consolidation pressure is missing.

Part geometry: Very flexible, undercuts are also possible. Typical strand diameters are at about 2 mm, which directly affects the possible printing resolution and possible radii [58].

Typical fields of application: Until now, there is no significant industrial application for 3D-printed FRP components.

Input materials: Fiber-reinforced (short or continuous fibers) thermoplastic strands or dry rovings combined with pure thermoplastic (for melt impregnation).

Typical quantities: Due to the low output rates, 3D printing is currently more suitable for small series.

To be considered: Methods for 3D printing of FRPs can be found almost exclusively in the field of research.

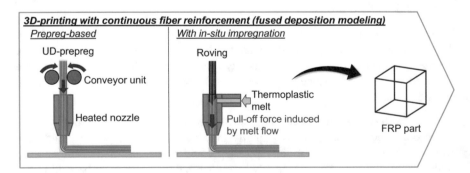

Fig. 4.39 Schematic illustration of the process "3D printing" with continuous fiber reinforcement

4.6.2.7 Tape Laying

Description: Tape laying is defined as the discontinuous placement of tapes onto a tool, variable in terms of direction and position. Here, the term tape refers to a semi-finished prepreg product with unidirectional fiber reinforcement. For the placement, e.g. a tape laying head mounted to an industrial robot or a gantry system can be used. Thermoset tapes provide inherent stickiness, the so-called tack, due to the uncured resin. This tack can be used for the placement on a tool or already placed tapes. After placement, the component must undergo a curing process under pressure and temperature, for example in an autoclave. For thermoplastic tapes (Fig. 4.40), the tape laying head comprises a heat source that melts the tape to be applied to the tapes that have already been placed. A consolidation roll presses the tape to the already placed tapes, so that a molecular diffusion can take place. At the same time, the roll also cools the tapes down below melting temperature again, to reach solidification. This way, consolidated components can be produced in situ; i.e. the final part can be directly manufactured in the tape laying process. Alternatively, tapes are only attached during tape laying and the full consolidation as well as the forming process take place in a subsequent thermoforming process [59–61].

Structure of the fiber reinforcement: Continuous fiber-reinforced, potentially with locally load-adapted fiber positioning and orientation. For the thermoplastic tape laying, the fiber volume content depends on the fiber volume content of the processed tapes, which is usually between 30 and 70% [19]. In thermoset tape laying, pressing out excess resin is possible, if an autoclave process is used. Hence, a higher fiber volume content higher than that of the initial tape is possible, typically between 60 and 70%.

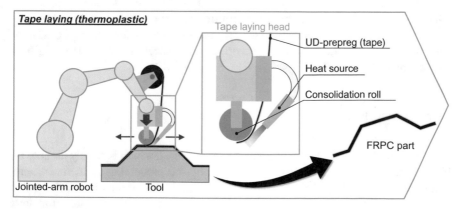

Fig. 4.40 Schematic illustration of the process "tape laying" (process variant with thermoplastic tapes)

Part quality: For thermoset tape laying, the subsequent autoclave process leads to very high component qualities with minimal porosity (<2%). In thermoplastic tape laying with in situ consolidation, process quality can also be very high. However, the process result strongly depends on the input quality of the tapes and a proper process design [62, 63]. In thermoplastic tape laying with subsequent consolidation a relatively high quality can also be achieved when applying high pressures.

Part geometry: Shell-like components are possible. Strong curvatures and small radii are however limited by the geometry of the tape laying head. Curved tape placement within the placement plane leads to defects (compression on the inner radius), caused by the lack of elasticity of the rovings. When using a gantry system, even very large components can be manufactured [59, 60].

Typical fields of application: In the aviation industry, thermoset tape laying is used for the production of fuselage, tail unit and wing structures. Thermoplastic tape laying is by far less common but in principle also suitable for shell components in aviation. Possible applications are also seen in the automotive sector.

Input materials: Continuous, unidirectional fiber-reinforced prepreg materials (thermoset and thermoplastic).

Typical quantities: The thermoset tape laying with subsequent processing in an autoclave is suitable for small and medium quantities. The thermoplastic tape laying combined with subsequent thermoforming is generally also suitable for large series. The thermoplastic tape laying with in situ consolidation is suitable for small series due to the flexibility and the low placement speeds.

To be considered: Contrary to flat semi-finished products, e.g. based on fabrics, near-net-shape placement of the final part is possible, which results in a significantly higher material efficiency. This is achieved on the one hand by reducing waste and on the other hand by providing the possibility of a locally load-adapted fiber reinforcement. Thereby, e.g. the wall thickness can be adapted to the local load instead of having a constant wall thickness. On the down side, the cycle time is relatively high, which results from the requirement of having to build the component tape by tape. In order to exploit the process advantages, tape laying can also be combined with fabric-reinforced thermoplastic sheet material, to reach a load-adapted design [64].

4.6.2.8 Autoclave

Description: An autoclave is a heatable pressure vessel, which is widely useable for FRP production. The classic variant is given by the processing of thermoset prepregs. As shown in Fig. 4.41, the prepreg laminate is placed on a tool and covered with supplementary materials, including release films, which allow for later demolding, a bleeder that absorbs excess polymer material, and finally a vacuum bag. Under the vacuum foil (in the area of the prepreg), a vacuum is generated. At

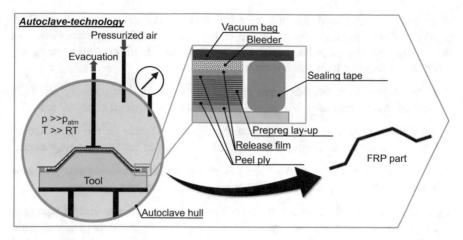

Fig. 4.41 Schematic illustration of the "autoclave technology" (process variant with thermoset prepregs)

the same time, the complete setup is placed in the pressurized and heated autoclave. The pressure leads to consolidation and squeeze-out of the excess resin, the heating leads to curing of the matrix [65].

Autoclave technology can also be used to process thermoplastic prepreg materials or laminates consisting of alternating fiber layers and polymer films. Typical process pressures are 6–7 bar for thermoset prepregs, with processing temperatures ranging between 100° and 200 °C. For thermoplastics, the required pressures are around 10 bar, and especially for high-performance thermoplastics temperatures of sometimes more than 400 °C must be reached. In order to meet these high requirements, autoclaves have been built that reach pressures of 70 bar and temperatures of 650 °C at a diameter of 3.5 m. In general, the autoclave technology is therefore very flexible concerning the materials that can be processed [20, 66].

Structure of the fiber reinforcement: Basically, all types of fiber materials can be processed. However, given the enormous effort, continuous fiber reinforcements are the standard case when it comes to industrial application. Autoclaves are often also used for prototyping. The manual preparation, which is still standard, also allows the production of locally load-adapted structures. Thermoset tape laying is an established preliminary process, especially in aerospace industry. Due to the high pressure, depending on the structure of the fiber reinforcement, a very high fiber volume content of 60–70% can be achieved.

Part quality: Due to the high pressures, the laminate quality achieved in the autoclave is so high that it is usually seen as the benchmark for all other processes, both with regard to the pore content and with regard to the fiber volume content. With fully automated tape laying processes, also a high accuracy in terms of fiber orientation can be reached. If a high surface quality is required on both sides, an upper tool must be used instead of a vacuum bag or an appropriately formed steel sheet must be integrated into the vacuum setup (under the vacuum bag). With regard to dimensional accuracy, the so-called spring-in effect (deformation after demolding) must be taken into account, which is quite difficult to predict, especially when an anisotropic fiber reinforcement is given.

Part geometry: Very flexible when manual lay-up is applied; partially and fully automated lay-up processes are useable for mostly shell-shaped components.

Typical fields of application: High-performance components for the sports sector, automotive small series (high-priced segment), as well as aerospace components.

Input materials: Continuous fiber-reinforced thermoset and thermoplastic prepreg materials or alternating stacks of dry fiber structures and polymer films.

Typical quantities: Due to the time-intense preparation of the setup (especially vacuum bagging) and the long cycle times, the autoclave technology is only suitable for small series.

To be considered: The autoclave technology is a classical process and offers enormous flexibility. When designing FRP parts, many design engineers have the autoclave process with all its possibilities in mind. However, components are then often not suitable for cost-efficient production with other process technologies. As the autoclave technology is only suitable for small series, this can lead to problems in the development of FRP components.

4.6.2.9 Prepreg Compression Molding

Description: The first step of this process is to process thermoset prepreg materials into a near-net-shape preform, whereby the material-inherent tack is exploited. This preform is then placed in a temperature-controlled mold (Fig. 4.42), whereby the heating initiates the curing of the thermoset. Compared to alternative technologies, such as autoclave technology, relatively fast curing can be reached due to the efficiency of the conductive heating [67].

Structure of the fiber reinforcement: Continuous fiber-reinforced, depending on the preforming process also locally load-adapted. A high fiber volume content (>60%) can be reached, due to the use of pre-impregnated prepregs. However, since there is no bleeder, the maximum fiber volume content is slightly lower than compared to autoclave technology.

Part quality: A double-sided tool ensures high surface quality on both sides [67]. Laminate qualities at autoclave level are possible [68].

Part geometry: The component must provide a shape that leads to uniform tool pressure, which is why shell-shaped components are particularly suitable. However, with a pressurized inner hose, hollow components are also possible, e.g. bicycle forks or handlebars [68].

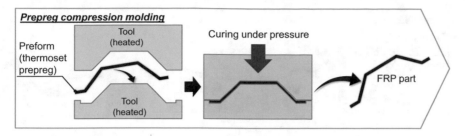

Fig. 4.42 Schematic illustration of the process "prepreg compression molding"

Typical fields of application: Shell-shaped components for small series in the automotive industry (high-priced segment) and bicycle components [67, 68].

Input materials: Thermoset prepregs with unidirectional or textile reinforcement.

Typical quantities: Compared to autoclave technology, significantly higher output can be achieved. This makes the process suitable for small to medium quantities. With appropriate design and the use of fast curing prepregs annual production of more than 80,000 parts per tool is possible [67].

To be considered: The process is in competition with resin injection processes, where impregnation takes place directly in the process and thus allows one to avoid expensive semi-finished products. However, due to the mostly complex flow patterns in near-net-shape preforms, process robustness is then also reduced. Furthermore, thermoforming with thermoplastic prepregs is an alternative. It allows for even shorter cycle times, as no curing is required, but also requires equipment that is more expensive.

4.6.2.10 Thermoforming

Description: Thermoforming is used for the processing of flat thermoplastic prepregs, usually with fabric or mat reinforcement (also referred to as organo sheets). As shown in Fig. 4.43, these prepregs are first heated above melting temperature (e.g. by an infrared radiator field). The prepregs are then transferred to a forming tool, which is heated below melting temperature. During the pressing process the component is formed and the thermoplastic is cooled down and solidifies [69].

Structure of the fiber reinforcement: Continuous fiber-reinforced, typically by fabrics or mats. Furthermore, locally load-adapted sheets, material-efficiently produced by thermoplastic tape laying are becoming increasingly popular. The fiber volume content of fabric-reinforced organo sheets is typically 45–55%.

Part quality: A good laminate quality can be reached; yet, strong surface waviness results from the volume shrinkage caused by the relatively large temperature differences between forming and room temperature [70, 71]. Regarding the

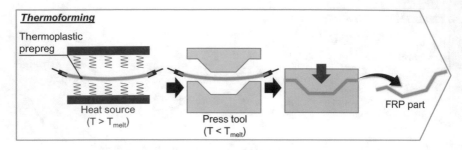

Fig. 4.43 Schematic illustration of the process "thermoforming"

dimensional accuracy, the so-called spring-in effect must be taken into account, a particularly important factor in the case of anisotropic fiber reinforcement. This deformation after demolding is difficult to predict. Another critical factor is the risk of wrinkling during forming [72].

Part geometry: The pressing process enables shell-shaped structures. The component size practically has no lower limit, but smaller components require corresponding production systems. The upper limit results from the required press and press tool that become more and more complex with increasing size.

Typical fields of application: Shell-shaped components in automotive engineering (for example seat shells [73]), increasingly also structural components for aviation (e.g. clips connecting frame, stringer and fuselage skin in the Airbus A350 [74, 75]) and sports products (skis/snowboards, protective helmets, etc.).

Input materials: fabric-reinforced thermoplastic sheets (so-called organo sheets) or mat-reinforced thermoplastics. Glass fibers but also carbon fibers are commonly used.

Typical quantities: As fully impregnated semi-finished products are used, the cycle time for the actual thermoforming process in the press is quite short. This makes the process fully suitable for large-scale production. On the other hand, presses and press tools require a relatively high investment, which is why the process is only suitable for medium series sizes, if the press capacity can be exploited by other products.

To be considered: Thermoforming is ideal for the production of sandwich laminates with a TP-FRP top layer on both sides and a core consisting of a foam, for example. The complete sandwich structure can be heated and formed together [20].

4.6.2.11 Hand Lay-Up

Description: In this process, a tool is covered alternately with a fiber structure (textile or mat) and a thermoset resin (see Fig. 4.44). Auxiliary equipment such as

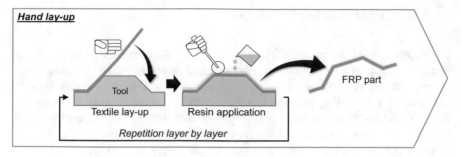

Fig. 4.44 Schematic illustration of the process "hand lay-up"

rollers are used to vent the laminate, improve the impregnation and compact the laminate [19].

Structure of the fiber reinforcement: Typically, random fiber mats are used, resulting in a random fiber orientation. However, textile fabrics can also be used. The achievable fiber volume content for mat laminates is about 15–20% and for fabrics about 40–50%. However, to reach the high fiber volume contents, curing under vacuum bagging is required [52].

Part quality: During hand lay-up, a large amount of air is inevitably pushed into the component, which later leads to a high porosity in the laminate. Furthermore, a smooth surface is only achieved on the side facing the tool. The side facing away from the tool can be improved by applying a vacuum bagging with integrated metal sheets for final curing. The dimensional accuracy is quite low due to the manual work and missing compaction force. It mainly depends on the experience of the skilled worker.

Part geometry: The high fraction of manual work, in combination with the relatively low quality requirements, allows for relatively complex structures. However, due to the use of areal textiles shell-shaped components are the standard case.

Typical fields of application: containers, ship hulls, prototyping

Input materials: Areal textiles, especially glass fiber mats, but also fabrics in combination with thermoset resin systems.

Typical quantities: Due to the manual effort and the low investment, hand lay-up is suitable for quantities up to 1000 parts per year [52].

To be considered: Given the low investment costs, hand lay-up is a good starting point for prototyping, especially for components which are intended to be produced by liquid composite molding. However, the qualitative differences in comparison with these procedures must be taken into account.

4.6.2.12   Liquid Composite Molding

Description: Liquid composite molding refers to all processes in which a fiber structure is impregnated with a low viscosity and therefore usually thermoset resin system, driven by positive or negative pressure. The fiber structure is typically prepared in a preceding preforming process. In this process, near-net-shape preforms are produced out of semi-finished products such as rovings or textiles. In the preform, the fiber position and orientation already corresponds to the intended position and orientation in the final component. Today the process group is very diverse, with various process variants. Only the most important main groups are presented in the following section.

In the vacuum infusion process, shown in Fig. 4.45, the preform is positioned on a tool and covered with a vacuum bag. The impregnation with the resin then takes place driven by the pressure difference between atmospheric pressure, outside the bag, and the vacuum in the preform area [20].

The resin transfer molding (RTM) process is shown in Fig. 4.46. Here, the preform is placed in a closed mold and the impregnation is driven by overpressure, created by a pressure vessel filled with resin or an injection system. An additional vacuum can help to prevent air inclusions. During standard RTM, injection pressures of up to 20 bar are achieved. Injection pressures in high-pressure RTM are often even beyond 200 bar and intend to lead to a faster impregnation [20, 76].

Many components manufactured by LCM processes are shell-shaped and have a very high aspect ratio, i.e. the surface area is many times greater than the dimension in the thickness direction. To reduce the impregnation time, there are therefore also process variants that allow impregnation in the thickness direction. This includes wet pressing (Fig. 4.47), for example. Here the resin system is first distributed onto the preform and the actual impregnation is then carried out by tool closing [20].

In addition to these process variants there are several other variants. Despite the different impregnation strategies, all these processes have in common that the flow distance of the matrix material within the preform is relatively long. This is only possible if the viscosity of the resin system is correspondingly low (<1000 mPa s) and the permeability of the textile structure is accordingly high.

Structure of the fiber reinforcement: Continuous fiber-reinforced, often based on textiles (woven and non-crimp fabrics). The fiber volume content is typically between the 45 and 60%, depending on the process and fiber structure.

Part quality: In processes with closed tooling a good surface quality can be reached. With single-sided tools, e.g. as for vacuum infusion, the wall thickness is uneven. A good laminate quality is reachable, with a good venting strategy and especially when vacuum is used.

Part geometry: Processes such as vacuum infusion, in combination with room-curing resin systems, allow the production of very large components with large wall thicknesses, such as ship hulls or wind turbine blades. Other processes,

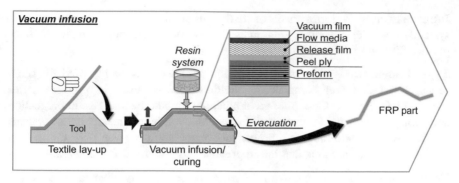

Fig. 4.45 Schematic illustration of the process "vacuum infusion"

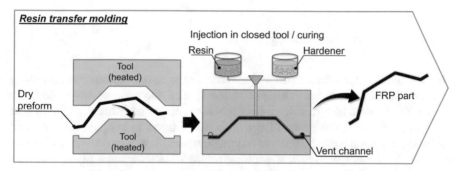

Fig. 4.46 Schematic illustration of the process "resin transfer molding"

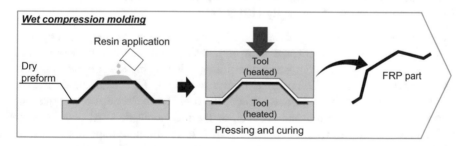

Fig. 4.47 Schematic illustration of the process "wet compression molding"

like RTM, also allow the production of small, relatively complex components such as brake levers for bicycles [77].

Typical fields of application: Structural components for automobiles (e.g. i-series of BMW), aerospace components (e.g. pressure dome for Airbus A350 XWB [78]), ship hulls, blades for wind turbines, structural elements in mechanical engineering.

Input materials: Preforms made of textiles in combination with thermoset resin systems. Possible, but rather seldom is the use of in situ polymerizing thermoplastics [79] and vitrimers [80].

Typical quantities: While vacuum infusion processes are mainly used for small series, RTM and wet compression molding can be used to produce series sizes >100,000 ppa as these offer great potential for automation. Fast impregnation and curing are essential for large series sizes. Modern, highly reactive resin systems can cure within one minute, if a proper tool temperature is given [79]. Using such systems, however, requires that the impregnation is finalized fast enough.

To be considered: The group of LCM processes is very diverse. The design possibilities, the economically producible series sizes, as well as the quality of the components can differ very much between the different variants.

4.6.2.13 Compression Molding

Description: Compounds for compression molding are prepregs made of short/long fibers and a matrix material. When pressure is applied, the material flows within the cavity. With regard to thermosets, sheet molding compound (SMC) is of particular interest. SMCs typically contain large amounts of fillers in addition to fibers and resin. The intention behind the fillers is, among others, to reduce material shrinkage, costs or the density (however, depending on the objective, some fillers may also increase density). Thermoset molding compounds are pressed in a heated tool (Fig. 4.48). The tool contact therefore initially reduces the viscosity, but also accelerates the curing reaction. As a result of the flow process, the compound is formed into the shape of the cavity [19].

Long-fiber-reinforced thermoplastics (LFT) can be seen as the thermoplastic equivalent to SMC. LFT consists of a granulated, unidirectional fiber-reinforced thermoplastic strand. For compression molding, it is first melted and plasticized in an extruder (Fig. 4.49). Compressed in the mold, the LFT compound fills the cavity. In contrast to the thermoset variant, the mold temperature is lower than the temperature of the semi-finished product, as the thermoplastic must cool down to solidify, in order for the part to be demolded. A formerly quite popular alternative, glass mat-reinforced thermoplastic, has been largely replaced by SMC and LFT [19].

Structure of the fiber reinforcement: Short/long fiber-reinforced (fiber length typically between 25–50 mm for SMC and 10–25 mm for LFT) with random fiber orientation. However, local fiber alignment in the flow direction is possible. The fiber content for molding compounds is usually given in fiber weight percentage and for SMC it is typically at about 20–30% by weight (corresponding to about 14–22% by volume). For structural applications up to 50% by weight are possible. With LFT granules, the fiber content is usually between 20 and 50% by weight

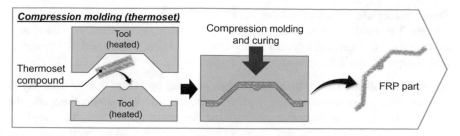

Fig. 4.48 Schematic illustration of the process "compression molding" (process variant with thermoset compound)"

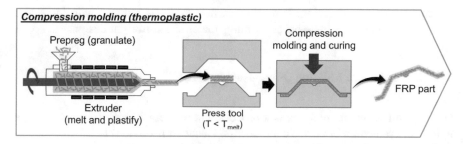

Fig. 4.49 Schematic illustration of the process "compression molding" (process variant with thermoplastic granulate)

(about 10–30% by volume), but up to 80% by weight is possible. Even higher performance is offered by newer high-performance SMCs, in which carbon fibers with a length of up to 50 mm and a fiber content exceeding 60% by weight (equivalent to 50% by volume) are embedded in an epoxy matrix (without fillers) [13, 20, 81–83].

Part quality: SMC offers the possibility of excellent surface quality and dimensional accuracy, when appropriate fillers are used and the fiber weight content is below 30%. Other than with textile-based FRPs, there is no danger of a print-through of a textile structure [13]. The surface quality reached with LFT is usually not suitable for visible components [84].

Part geometry: The flow process allows manufacturing components with rib structures, which is not possible with continuously fiber-reinforced FRPs. The flowability of the compound is strongly influenced by the fiber length, the fiber volume content and, if present, the fillers.

Typical fields of application: Semi-structural components, such as tailgates. SMC made of glass fibers is also an excellent insulator and is therefore widely used, e.g. for fuse boxes.

Input materials: Short/long fiber-reinforced prepregs (thermoplastic and thermoset). The fiber material is usually (not exclusively) glass fiber, typical matrix materials are unsaturated polyester resins and vinyl esters with regard to thermosets and polypropylene and polyamide with regard to thermoplastics [19].

Typical quantities: Due to the very efficient pressing processes and the usability of the compounds as standard semi-finished products for various components, the compression molding processes are suitable for mass production. The high investment costs for the press and the tools make an economic production of small and medium series difficult and require that the capacity utilization of the press is maximized by production of other components.

To be considered: Compression molding offers outstanding possibilities for structural lightweight design. However, as flowability is reduced with increasing fiber length, there is a trade-off concerning material lightweight design.

4.6.2.14   Injection Molding (Fiber-Reinforced)

Description: Injection molding is a process with extremely high relevance for the industrial production of plastic components. Hence, it is obvious to use this process for the production of FRP components. In the classic thermoplastic variant a granulate of short or long fiber-reinforced thermoplastic strand is melted and plasticized in an extruder (Fig. 4.50). Thereby, the screw in the extruder pushes itself backwards and is then finally pushed forward, which injects the material into the cavity and fills it. A thermoset variant is also available, referred to as bulk molding compound [19, 85, 86].

Structure of the fiber reinforcement: Short/long fiber-reinforced (medium fiber length typically less than 3 mm for TP injection molding [19] and less than 12 mm for bulk molding compound (BMC) [20]) with largely random arrangement. A local fiber alignment induced by the flow is possible [87]. The fiber weight content for thermoplastic injection molding is 10–60% (about 5–40% by volume) [86] , and for BMC, it is between 15 and 30% (about 10–20% by volume).

Part quality: Bulk molding compound allows excellent surface qualities [19]. For thermoplastics, possible quality problems induced by moisture absorption [88] have to be taken into account.

Part geometry: The comparatively high flowability allows the production of components with complex rib structures. The flowability is determined by fiber volume content and fiber length. The component size is limited by the required the press tool and closing unit as well as the possible flow paths.

Typical fields of application: Geometrically complex but non-structural elements. Often relatively small, e.g. connecting components, but also larger components such as bumpers.

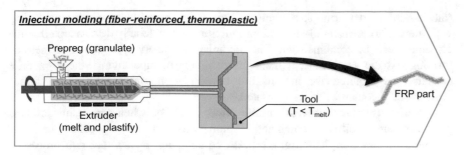

Fig. 4.50 Schematic illustration of the process "injection molding" (process variant with fiber-reinforced thermoplastic granulate)

Input materials: Short fiber-reinforced granulates (thermoplastic and thermoset). The fiber material is usually (not exclusively) glass fiber, typical matrix materials are unsaturated polyester resins as thermosets and polypropylene and polyamide as thermoplastics [19].

Typical quantities: Injection molding processes are industrially established, and due to the comparatively short cycle times, these processes are suitable for mass production.

To be considered: Injection molding processes offers outstanding possibilities for structural lightweight design. Yet, as flowability is reduced with increasing fiber length, there is a trade-off concerning material lightweight design.

4.6.2.15 Process Hybridization

Each of the processes presented in the previous section offers specific advantages and disadvantages. The combination of two or more processes can potentially combine the advantages while disadvantages are compensated. There can be different intentions behind the development of such process combinations:

1. The endeavor to use material-efficient manufacturing processes for semi-finished products
2. The improvement of the component's mechanical performance through locally load-adapted reinforcement
3. Increasing the potential for lightweight design by combining material lightweight design (continuous fiber reinforcement) and structural lightweight design (short/long fiber reinforcement) in a single component

In general, **two approaches to process hybridization** can be distinguished.

A **substitutive process combination** is given, when the result of one process is used as input for another process and thereby substitutes the usually used semi-finished product. For example, a preform made by thermoplastic tape laying, can be processed in a thermoform process, instead of the usually used organo sheet (fabric-reinforced thermoplastic sheet). This way, the advantages of tape laying with regard to a material-efficient production and a locally load-adapted reinforcement can be combined with the advantageous short cycle times of thermoforming. At the same time, a main disadvantage of tape laying, the high time consumption required for in situ full consolidation and placement of three-dimensional geometries is compensated.

A **complementary process combination** is given, when the results of two processes are additively combined. Again, there are two possible variants. A **simultaneous complementary** process combination means that two processes are combined into one common process. For example, the thermoforming of an organo sheet can be combined with an injection molding process, so that for example a stiffening structure out of pure polymer is injected onto the organo sheet. A **sequential complementary** combination of processes means that the result of one process that already represents a FRP component is further processed in a second process. For example, an organo sheet formed by thermoforming can be further processed by means of a thermoplastic tape-laying process, which adds a local reinforcement. Complementary process combinations therefore usually result in hybrid materials comprising two different structures of fiber reinforcement.

Beyond these two basic combination possibilities, of course hybrid process can also be both, substituting and complementary. Referring to the examples given above, e.g. a preform made by thermoplastic tape laying could be processed in a combined thermoforming/injection molding process.

Overall, this results in various possible combinations. Table 4.15 provides an overview. Due to the rapidly progressing development and the numerous combinations, it is claimed neither that this overview is complete, nor that all the examples have already been developed or even industrially applied. The intention is to show which combinations are generally possible and to convey to the reader the way of thinking behind the concept of process hybridization.

4.6.2.16 Industrial Relevance of the Processes

Finally, the industrial distribution of the different processes for the production of components made of GFRP and CFRP is considered. Since the fiber material and the structure of the fiber reinforcement have already been determined, this consideration can be a useful indicator for the process selection, even if, of course, no generally applicable guideline can be derived from these statistics. For example, it is not impossible, but very unlikely that prepreg-autoclave is the optimum process for a GFRP component. This can be seen from the fact that this process has almost no industrial relevance for GFRPs.

Figure 4.51 shows the production quantities of GFRPs, sorted by different process groups. It can be clearly seen that infusion processes and short fiber-reinforced molding compounds are widely used. In addition, open processes still account for an enormous share. On the other hand, processes for the production of high-quality components in small series, such as the autoclave technology, have

Table 4.15 Examples of process combinations

	Tape laying	Compression molding	Injection molding
Thermoforming	S^a: Preform manufacturing by tape laying K_{seq}^b: Local reinforcement of an organo sheet before or after thermoforming	K_{sim}^c: Simultaneous forming and compression molding of organo sheets and thermoplastic compound, respectively	K_{sim}: Simultaneous forming of organo sheets and injection molding of thermoplastic injection compound
Liquid composite molding	S: Preforming via Dry Fiber Placement (Tape placement with polymer-bindered rovings) K_{seq}: Local preform reinforcement with dry rovings or thermoset tapes prior to impregnation	K_{sim}: Simultaneous compression molding of thermoset compound and impregnation of a dry fiber structure placed in the press tool with the resin of the compound	K_{sim}: Simultaneous injection molding of a thermoset compound and impregnation of a dry fiber structure placed in the injection tool by the resin from the compound
Prepreg compression molding	S: Preform manufacturing by tape laying K_{seq}: Local reinforcement of textile prepregs prior to pressing	K_{sim}: Simultaneous compression molding of a compound and a textile prepreg	K_{sim}: Simultaneous prepreg compression molding and injection molding of thermoset compound
Tape laying		K_{seq}: Local UD-reinforcement after the compression molding (thermoplastic) K_{seq}: Manufacturing of UD-reinforced inlay for the compression molding (thermoplastic)	K_{seq}: Local UD-reinforcement after injection molding (thermoplastic) K_{seq}: Manufacturing of UD-reinforced inlays for the injection molding (thermoplastic)

[a]S: Substituting process combination
[b]K_{seq}: Sequential, complementary process combination
[c]K_{sim}: Simultaneous, complementary process combination

only marginal industrial relevance. Short glass fiber-reinforced thermoplastics with fiber lengths <2 mm (injection molding) are not included in the overview. The European market volume for these materials was as big as 1470 kt in 2017 and was thus bigger than the market for the other processes together.

For the production of CFRPs, production technologies achieving high component quality, especially prepreg-based ones (Fig. 4.52), are dominant. In addition, processes with high material efficiency and high accuracy of fiber orientation, such as pultrusion and winding are of high relevance. On the other side, pressing processes for short fiber materials, where the fiber properties are not optimally exploited, are less common.

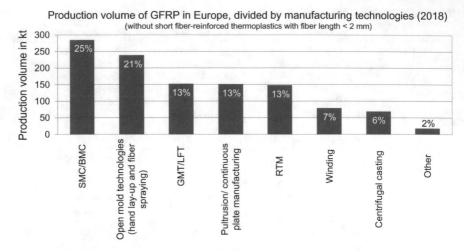

Fig. 4.51 Production volume of GFRPs in Europe, divided by manufacturing technologies, data from [89]

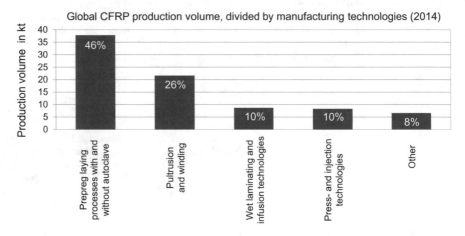

Fig. 4.52 Global CFRPs production volume, divided by manufacturing technologies (2014), data from [90]

4.6.3   Process Selection

Given the large variety of processes, selecting a specific process for the component to be developed is not a trivial task. In addition, due to the interdependencies of the process with the material and the design, it is obvious that this task should be jointly addressed by the different departments; i.e. IPD should be applied. Of course, the expertise of the team member from the manufacturing department is essential.

Based on some simple considerations, the number of possible processes can already be strongly reduced. Figure 4.53 shows a simple approach for these first considerations. The main criteria are the preliminary component geometry and the structure of the fiber reinforcement, which have already been determined. Some processes are very efficient in terms of both cycle times and material use. These processes will usually be used when applicable. This typically includes all the continuous processes, when it comes to the production of uniaxial profiles,[6] such as tubes. If the component is rotationally symmetric but not uniaxial (e.g. pressure vessels), continuous methods are not applicable and the winding technology is the method of choice. If the component's geometry is shell-like, the size of the component must be considered as a further criterion. For components <3 m^2 different pressing technologies can be used. Beyond 3 m^2, cost-efficiency of pressing technologies will rapidly decrease. The value 3 m^2 is an approximate value. Finally, there are structural components, which are not shell like and can therefore not be manufactured by forming a 2D semi-finished product. After considering the geometrical boundary conditions, the number of suitable processes can be further boiled down by taking into account the structure of the fiber reinforcement.

After these considerations, in many cases several generally applicable manufacturing processes will remain. For example, the thread lever, considered in the previous chapters, is a small shell-like part. Given the textile reinforcement, generally thermoforming, liquid composite molding as well as prepreg compression molding and hand lay-up are possibly suitable processes. For the final selection, therefore the technical and economical boundary conditions can be considered as follows.

- Comparison of the **component geometry** with the design possibilities of the processes.
- Comparison of the required mechanical performance with the typically achievable **material characteristics** of the processes in combination with the selected fiber material and the selected structure of the fiber reinforcement.
- Comparison of the planned **annual quantity** with the typical areas of application of the processes.
- Comparison of the **quality** requirements with the corresponding process characteristics.
- Comparison of the applicable processes with the pre-existing **expertise and facilities** of the company that commissions the product development.

For these comparisons, the estimates of the individual processes made in Sect. 4.6.2 can be used. In addition, Table 4.16 lists approximate values that enable a comparison of the processes with regard to the most important criteria. The processes are evaluated by assigning a numerical value between 1 and 5, whereby the legend contained in the table defines the meaning of the numerical values. It

---

[6]Uniaxial profiles are defined by a constant cross section which is extruded one dimensionally (geometrically). An I-beam is a classic example.

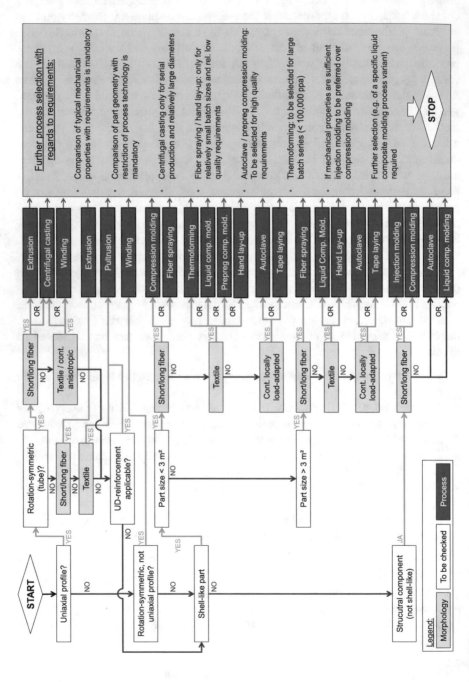

Fig. 4.53 Selection procedure for manufacturing processes

Table 4.16 Comparison of different manufacturing processes

Criteria / Processes	Part size	Part complexity	Part thickness	Annual output quantity	Accuracy of fiber orientation	Material efficiency	Investment	Dimensional accuracy	Laminate quality	Surface quality	Usable matrix polymer*
Fiber spraying	3-5	3-4	1-4	1-2	1	4	1	1	1	1	TS
Centrifugal casting	2-4	1	1-5	2-3	1-4	5	3	2	5	4-5	TS
Winding	2-5	1-2	1-5	2-3	4	5	3	4	4	2-4	TS
Pultrusion	2-5	1-2	1-4	3-5	4	5	3	5	4	3	TS
Extrusion	1-2	1-2	1-3	3-5	1	5	3	5	3	4-5	TP
3D-printing (continuously reinforced)	1-2	3-4	1-2	1	3-4	4	1	3	2	1	TP
Tape laying	2-4	1-3	1-3	1-3	4	4	5	3	4	2-3	TS/TP**
Autoclave	2-5	3-4	1-4	1-2	5	3	4	4	5	4-5	TS/TP
Prepreg compression molding	2-3	1-3	1-2	2-4	3-4	2	2	4	4	4-5	TS
Thermoforming	2-3	1-3	1-2	4-5	3-4	2	5	3	3	1-3	TP
Hand lay-up	3-5	2-4	1-5	1-2	3	3	1	2	1	3-5	TS
Liquid composite molding	2-5	1-3	1-5	1-5	3-4	2	1-5	3-5	1-5	3-5	TS
Compression molding	2-3	2-5	1-3	4-5	1	5	5	5	3	4-5	TS/TP
Injection molding	1-2	3-5	1-3	4-5	1	4	4	5	4	4-5	TP/TS

Legend:

Part size	1:	< 0.01 m²	→	5:	> 10 m²
Part complexity	1:	Strictly limited	→	5:	Large freedom of design
Part thickness	1:	< 5 mm	→	5:	> 50 mm
Annual production quantity	1:	< 1000	→	5:	> 100.000
Accuracy of fiber orientation	1:	Only short-fiber, random orientation	→	5:	Unidirectional, locally load-adapted
Material efficiency	1:	MIR***>130%	→	5:	MIR < 105%
Investment	1:	< 25,000 €	→	5:	> 1,000,000 €
Dimensional accuracy	1:	Deviation from nominal to actual value > 5%	→	5:	Deviation from nominal to actual value < 1%
Laminate quality	1:	Low, high porosity	→	5:	High, Porosity < 2%
Surface quality (single-side)	1:	Low, irregular thickness, high waviness and or roughness	→	5:	High, uniform thickness, high gloss

*TS = Thermoset, TP = Thermoplastic; the first named is always the standard case
**TP is only the standard case, when considering tape laying as single process. When combined with autoclave, TS is the standard case
***MIR = material input rate

should be noted that the values reflect the typical application area, which does not mean that the application outside this area is technically impossible. However, often it would not be cost-efficient and/or cause an increased development risk and expense.

The presented approach simplifies the selection methodology very much and serves to provide initial assistance in the selection of processes. Of course it will in many cases over-simplify, as very specific advantages and disadvantages of the processes, which may be relevant for a specific application, are not taken into account. For example, using thermoforming to manufacture thermoplastic FRP, allows to subsequently press in bearings. This is, e.g. highly interesting for the sewing machine thread lever [91]. This shows again clearly that a team member with appropriate expertise must be involved. However, with the help of the basic knowledge imparted here, the other team members as well can take into account manufacturing aspects at an early stage.

As explained in Sect. 4.6.2, most of the processes mentioned above represent process groups consisting of several process variants. Due to the intense research and development activities concerning manufacturing technologies for FRPs, the number of process variants continues to grow. Thereby, the individual process variants can strongly differ regarding the process strategy and the process result. The evaluations shown in Table 4.16 refer to the entire process group. Which specific process variant actually meets the special requirements must therefore be examined in detail. In some cases, the selection can be made relatively easy, by taking into account some boundary conditions. For example, of all the liquid composite molding processes, those based on vacuum infusion would be almost exclusively suitable for the low-volume production of a 10 m long hull of a sports yacht. Due to the low pressure, light and relatively inexpensive tools can be used, whereas RTM tools can easily weigh many tons for components with just a few square meters in surface area. A tool for the 10 m hull would not be cost-efficient for small series production. However, there are different vacuum infusion processes that differ in details, such as the two processes Seemann Composites Resin Infusion Molding Process (SCRIMP) and Controlled Atmospheric Pressure Resin Infusion (CAPRI). Estimating the effects on the process results requires a lot of experience. Sometimes a literature review can help, which for example reveals corresponding experimental or theoretical comparative studies on a reference part [92, 93] or even selection methods [94]. If no final decision can be made at this point, the most promising alternatives can all be further considered, to then select based on the techno-economic evaluation, taking into account the results of the prototype tests. Alternatively, only one alternative is further elaborated to eventually see if it is suitable. In general, the following applies: To minimize the development risk, industrially established variants should be preferred over newer variants. In addition, possibly existing intellectual property rights should be checked.

4.7 Decision for Thermoset or Thermoplastic Matrix and Evaluation of the Material Selection

The decision between thermoset and thermoplastic has far-reaching consequences for the development of the design and production concept. Therefore, this decision should be made as early as possible, in order to enable the simultaneous development in the different departments. For this reason, the decision for the polymer class is separated from the decision for a specific polymer and will take place at this point of the development.

In some cases, the process selected in the previous step already defines whether the matrix polymer can be a thermoplastic or a thermoset. For example, in liquid composite molding processes, thermosets are almost exclusively used and in the case of thermoforming, almost exclusively reinforced thermoplastics are used.[7] Other methods generally allow the processing of both polymer classes. However, here too, one polymer class can usually be seen as the standard case. In the last column of Table 4.16 the useable polymer classes are defined for different processes, whereas the first named is always the standard case. Assuming that initially the standard case should be chosen, the selected process directly determines the polymer class to be used.

In order to understand the consequences of this decision, the differences between the polymer classes are considered in the following, starting with a look at the molecular structure (Fig. 4.54).

A **thermoset** resin system cures to a cross-linked three-dimensional molecular network. This chemical process is irreversible—once cured, dissolution of the network is no longer possible without destruction.

Concerning the **thermoplastics**, two basic categories can be distinguished. **Amorphous** thermoplastics are made up of long, randomly intertwined molecule chains. When heated, the molecule chains can slide off from each other, so that thermoplastics can be formed and melted. When cooled down, the structure solidifies again; i.e. the process is reversible. In **semi-crystalline** thermoplastics, the molecular chains locally arrange themselves in regular crystalline structures. This crystallite formation is also reversible. In amorphous thermoplastics, a mobility of the molecule chains can be achieved by heating.

These structural differences affect the polymer behavior in many ways, concerning both processing and application. First, the mechanical property changes under the influence of temperature changes, referred to as the deformation behavior. The deformation behavior is often crucial for the application. The deformation behavior for thermosets as well as amorphous and semi-crystalline thermoplastics is shown in Fig. 4.55. In this case, the change in the dynamic shear modulus is shown

---

[7]Some systems do not match this clear distinction. For example, in situ polymerizing thermoplastics [95–97] can be processed in liquid composite molding and vitrimers [80] in forming processes. However, due to their relatively low prevalence, these will not be discussed in detail here.

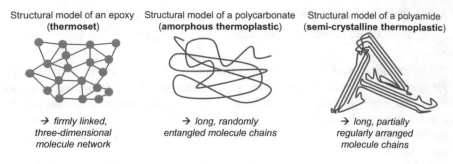

Structural model of an epoxy   Structural model of a polycarbonate   Structural model of a polyamide
(thermoset)   **(amorphous thermoplastic)**   **(semi-crystalline thermoplastic)**

→ *firmly linked,*   → *long, randomly*   → *long, partially*
three-dimensional   *entangled molecule chains*   *regularly arranged*
molecule network   *molecule chains*

Fig. 4.54 Comparison of structure models of thermosets as well as amorphous and semi-crystalline thermoplastics

as a function of the temperature. However, the curve looks similar, e.g. when the tensile or flexural modulus is considered.

The top diagram shows the deformation behavior of a cross-linked thermoset. Up to the so-called glass transition temperature, the behavior is energy elastic. This is the typical area of application for thermosets; here they have a high strength but are also relatively brittle, which is unfavorable with regard to the failure behavior. Above the glass transition temperature, not cross-linked areas soften, which, depending on the degree of cross-linking, results in a drop in properties. The higher the degree of cross-linking, the smaller the drop. Since the cross-linking is irreversible, the microstructure otherwise remains stable up to the point where decomposition begins. As a result, thermosets typically provide relatively high temperature stability.

For amorphous thermoplastics, the range of application is usually below the glass transition temperature. Above this temperature, the entangled molecule chains slide off each other, so that a deformation is possible. When the material is further heated eventually, a temperature is reached at which the material melts and a complete flow of the material is possible. At even higher temperatures, decomposition begins.

The application of semi-crystalline thermoplastics can be both, above and below the glass transition temperature, depending on the polymer. This is because above the glass transition temperature, the amorphous parts are thermo-elastic but the crystalline regions are still rigid. Application above the glass transition temperature results in a comparatively high toughness, which leads to a preferable failure behavior compared to thermosets. Only when the crystallite melting temperature is reached and the crystallite areas begin to melt, hot forming becomes possible. When all crystallite areas are melted, the flow temperature is reached and the thermoplastic can flow. Here too, a further increase in temperature eventually leads to decomposition.

While even cheap thermoset systems have a relatively high temperature resistance the use of thermoplastics can quickly become comparatively expensive, if high temperature requirements must be met. Low-cost thermoplastics have a

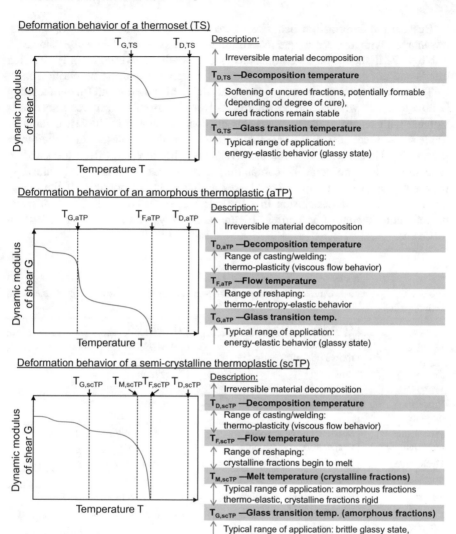

Fig. 4.55 Comparison of the deformation behavior of thermosets as well as amorphous and semi-crystalline thermoplastics. Adapted from [98, 99]

pronounced tendency to creep even at relatively low temperatures. Hence, if at this point the process selection has led to a thermoplastic variant and if relatively critical temperature requirements (>100 °C) are given, it should be carefully checked whether a suitable thermoplastic is available (see Sect. 5.3) and whether it is within a reasonable price range. As no detailed cost analyses are available at this point, a rough estimation must be made.

Besides the deformation behavior, there are several other advantages and disadvantages with respect to thermosets and thermoplastics. These are shown in Table 4.17. It should be noted that this is a generalized overview, only showing trends and not being universally valid for all thermosets and thermoplastics.

The task of the team members from the fields of materials and manufacturing is now to check whether the polymer class resulting from the process selection is suitable with regard to the requirements catalog. As shown in the last column of Table 4.16, there are processes that allow both thermoplastic and thermoset processing. Here a change is comparatively easy, but deviation from the standard case should only be done for compelling reasons. For processes, only allowing processing of one of the polymer classes a change is not possible. If the evaluation has led to the conclusion that the polymer class is unsuitable, a different process must therefore be chosen. Figure 4.53 shows alternatives for almost all processes.

Table 4.17 Qualitative comparison of processing-related characteristics of thermosets and thermoplastics

Category	Thermoset	Thermoplastic
Forming	Not meltable, virtually no formability after curing	Meltable and formable
Processing viscosity	Low	Very high
Processing pressures	Low	Very high
Processing temperatures	Systems curing at room temperature are available	High temperatures for forming and melting processes
Cycle times	Partially very long curing times	Fast consolidation/solidification
Shelf-life	Limited	Virtually unlimited
Solvents	Solvent vapors, volatile components in processing	Solvent-free processing
Welding	Not weldable	Weldable
Adhesive bonding	Very suitable	Critical, especially for semi-crystalline TP
Appearance	Good to very good surface quality	High shrinkage
Solvent resistance	High	High for semi-crystalline TP
Creepage	Low creep tendency	High creep tendency especially at elevated temperatures
Toughness	Relatively low	Relatively high

If, due to an inadequate data basis, no final decision for a polymer class can be made at this point, different alternative concepts should be further elaborated, so that the decision is postponed. The decision for a specific thermoset or thermoplastic system (e.g. epoxy resin or polyamide 66) is not made at this point, as further information is required, which can only be obtained from the elaboration of the drafts.

4.8 Definition of the Overall Draft

At this point, the results obtained so far must be aggregated into an overall draft. If, in the course of further elaboration, one of the aspects defined in the overall draft is changed, this should be discussed with all team members, to find out if this again requires further changes. For example, if the component thickness is strongly increased, it must be checked, if the chosen material and manufacturing process are still suitable.

It is advisable to prepare a list with all key figures to have a clear overview at all time. Table 4.18 shows an example of a simple list with some example values.

With the finalization of the overall draft(s), Milestone M2 (Fig. 2.1) is reached.

4.9 Identification of the Drafts to Be Further Elaborated

The procedure described in the previous sections will not necessarily allow the definition of a single solution that is most likely to be the techno-economically best option. On the contrary, at this point there will probably be several solutions that appear to be relatively similar, concerning their chances to be the best option. The final selection will ultimately be made after techno-economic evaluation, as the last

Table 4.18 Example for a simple list of the relevant characteristics of a full draft

Category	Property	Selection
Reinforcement	Fiber material	CF-AF-hybrid (lightweight, danger of impact)
	Structure of fiber reinforcement	Continuous anisotropic, textile-reinforced
Part geometry	Thickness	Varying between 2.5 and 3.0 mm
	Area	2.5 m^2, closed surface
	Form	Shell-like, double-curved, no undercuts
Manufacturing	Process group	Liquid composite molding
	Variant	Resin transfer molding
Matrix	Polymer type	Thermoset (process-induced)

step of the proposed procedure. However, the further steps up to this point will involve enormous effort. Hence, a resource-efficient development strategy should include mechanisms to reduce the number of drafts to be elaborated to a reasonable level. For this task, it has to be taken into account that the risk that eventually none of the elaborated drafts lead to satisfying results, increases with a decreasing number of elaborated alternative drafts. If this worst case occurs, one would have to start all over again with the elaboration, focusing on one of the initially not selected drafts. This would result in a loss of time. Accordingly, it is necessary to balance the risk of time loss on the one hand and the resources for draft elaboration on the other.

The risk that initially promising drafts will ultimately prove to be a "dead end" should not be underestimated, especially when FRPs are applied. In Sect. 1.6, it was explained that the relevance of IPD for FRPs mainly results from the rapid development of new materials and processes, which brings a particular complexity. Accordingly, when developing FRP components, it is quite common, e.g. that a possible solution is found which is potentially superior to the current state of the art, but with questionable feasibility. In order to minimize the development risk, it is therefore advisable to rely on proven technologies, especially if the development team has comparatively little experience with the development of FRP components and thus with corresponding risk assessments. However, due to the rapid development of FRP technologies, new approaches often bear enormous potential. Therefore, in more experienced development teams, an appropriate assessment of the risks, as well as the expected efforts for further elaboration should be part of every product development. The results must be compared with the customer's willingness to take risks, which again depends on numerous factors.

Methods for the assessment of developments risks and costs or for the assessment of the customer's willingness to take risks will not be detailed at this point (an overview is, e.g. provided by [100]). However, it is clear that balancing these aspects is critical for the successful solution of the development task. Even a subjective assessment of the development risk by experienced developers and a corresponding open discussion with the customer can already provide valuable insights. Many solutions can already be excluded by rather trivial considerations: E. g. a small company that wants to launch a new product within months will not be interested in a risky development of a new process technology requiring months of preliminary studies. No matter how elegant the solution is, the company will opt for a more conservative variant. Accordingly, at this point a meaningful selection of drafts is to be selected for further elaboration. For this, the overall procedure shown in Fig. 2.1 provides for a separate meeting under involvement of all team members and the customer (step 2.4). This should include comparison of the entire requirements catalog with the drafts, to ensure that the drafts selected for further elaboration can potentially meet all minimum requirements.

4.10 Questions for Self-Check

Below are some questions and tasks to help you reflect on the main contents of this section. The solutions can be found in Sect. 4.6.

R12. Glass fibers have a negative coefficient of thermal expansion—True or false?

R13. The stiffness and strength of carbon fibers is determined by the final process temperature in their manufacturing process—True or false?

R14. The elongation at break of thermoplastics is typically higher than that of thermosets—True or false?

R15. An effective reinforcement of a polymer with fibers is only is achieved when the elongation at break of the polymer is lower than that of the fiber—True or false?

R16. An effective reinforcement of a polymer with fibers is only achieved when the modulus of elasticity of the fibers in fiber direction is higher than that of the polymer—True or false?

R17. Briefly describe the molecular structure of thermosets as well as amorphous and semi-crystalline thermoplastics.

R18. Name one task each for fiber, matrix and interphase in a fiber-reinforced polymer.

R19. If a unidirectional fiber-reinforced polymer is loaded parallel to the fiber direction, a parallel connection of springs is a proper substitute model—True or false?

R20. A typical specimen for tensile tests made of unidirectional continuous carbon fiber-reinforced epoxy resin has a pronounced yield strength (high elongation at break)—True or false?

R21. Draft a diagram with the normalized properties from 0 to 1 on the vertical axis and the fiber length from 0 to 100 mm on the horizontal axis. Include the typical curves showing the corresponding correlation for strength, stiffness and impact strength of a component made of long glass fiber-reinforced polypropylene (fiber volume content 40%).

R22. Draft a diagram with the normalized strength on the vertical axis and the fiber angle on the horizontal axis. Include the typical curves showing the corresponding correlation for a unidirectional, a bidirectional ($[0°/90°]_s$) and a quasi-isotropic fiber-reinforced polymer.

R23. Define the terms "material, structural and system lightweight design."

R24. Name three advantages and three disadvantages of the differential and the integral design method.

R25. Name five types of deformation couplings. For one of these types give an exemplary laminate structure providing this coupling.

R26. With a cross-sectional area of 50 mm^2, a 1000 mm long bar of uni-directional fiber-reinforced polymer should have a maximum defor-mation of 2 mm at a tensile load of 4000 N. Calculate which stiffness is required for this. What is the minimum required fiber volume con-tent, when using a glass fiber with a stiffness of 72 GPa and with negligible stiffness of the matrix?

R27. Name three specific advantages and three specific disadvantages of a thermoset matrix.

R28. Name three specific advantages and three specific disadvantages of a thermoplastic matrix.

R29. Plot the typical deformation behavior of the material of a semi-crystalline thermoplastic over the temperature in a diagram. Mark the glass transition temperature, the range of thermoelasticity and the range of thermoplasticity.

R30. Define the terms impregnation, consolidation and solidification.

R31. Name four specific advantages and four specific disadvantages of fiber–plastic composites.

R32. With fiber materials, the smaller the diameter, the higher the mechanical properties observed. Name three effects that explain this observation.

Literature

1. Kelly, A., Davies, G.: The principles of the fiber reinforcement of metals. Metall. Rev. **10**(1), 1–77 (1965)
2. Courtney, T.H.: Mechanical Behavior of Materials, 2nd edn. Waveland Press, Illinois (2005)
3. Fu, S.-Y., Lauke, B., Mäder, E., Yue, C.-Y., Hu, X.: Tensile properties of short-glass-fiber- and short-carbon-fiber reinforced polypropylene composites. Compos. A Appl. Sci. Manuf. **31**(10), 1117–1125 (2000)
4. Thomason, J., Vlug, M., Schipper, G., Krikor, H.: Influence of fiber length and concentration on the properties of glass fiber reinforced polypropylene: Part 3. Strength and strain at failure. In: Composites Part A: Applied Science and Manufacturing, vol. 27 (11), pp. 1075–1084 (1996)
5. Pan, N.: Theoretical determination of the optimal fiber volume fraction and fiber–matrix property compatibility of short fiber composites. Polym. Compos. **14**(2), 85–93 (1993)
6. Batch, G.L., Cumiskey, S., Macosko, C.W.: Compaction of fiber reinforcements. Polym. Compos. **23**(3), 307–318 (2002)
7. Thomason, J.: The influence of fiber length and concentration on the properties of glass fiber reinforced polypropylene: 5. Injection moulded long and short fiber PP. In: Composites Part A: Applied Science and Manufacturing, vol. 33(12), pp. 1641–1652 (2002)

8. Verein Deutscher Ingenieure e.V.: VDI-Richtlinie 2014: Entwicklung von Bauteilen aus Faser-Kunststoff-Verbund (Blatt 1: Grundlagen, 1998, Blatt 2: Konzeption und Gestaltung, 1993, Blatt 3 Berechnungen, 2006) (2006)
9. Jones, R.M.: Mechanics of Composite Materials. CRC Press, Boca Raton (1999)
10. Montano, Z., Kühn, M., Daniele, E., Stüve, J.: Biege-Torsionskopplung an Rotorblättern für Windenergieanlagen. Lightweight Des. **11**(4), 46–51 (2018)
11. Bergmann, H.W.: Konstruktionsgrundlagen für Faserverbundbauteile. Springer, Berlin/ Heidelberg (2013)
12. Mittelstedt, C., Becker, W.: Strukturmechanik ebener Laminate. Technische Universität Darmstadt (2017)
13. Schürmann, H.: Konstruieren mit Faser-Kunststoff-Verbunden. Springer, Berlin (2007)
14. Hertel, H.: Leichtbau: Bauelemente, Bemessungen und Konstruktionen von Flugzeugen und anderen Leichtbauwerken. Springer, Berlin (2013)
15. Verein Deutscher Ingenieure e.V.: VDI 2221—Methodik zum Entwickeln und Konstruieren technischer Systeme und Produkte (1993)
16. Steinhilper, R., Rieg, F.: Handbuch Konstruktion. Carl Hanser Verlag GmbH Co. KG, Munich (2012)
17. Pahl, G., Beitz, W., Schulz, H.-J., Jarecki, U.: Pahl/Beitz Konstruktionslehre: Grundlagen erfolgreicher Produktentwicklung, Methoden und Anwendung. Springer-Verlag, Berlin/ Heidelberg (2013)
18. Lahr, R., Noll, T., Mitschang, P., Himmel, N.: Auslegung eines Fadenhebels und Entwicklung eines neuen Fertigungskonzepts für Formnestumformung bei thermoplastischen CFK-Bauteilen. Landshuter Leichtbaukolloquium 2005, Landshut, 24 (2005)
19. AVK-Industrievereinigung Verstärkte Kunststoffe (ed.): Handbuch Faserverbundkunststoffe/ Composites: Grundlagen, Verarbeitung, Anwendungen, 3. Auflage. Springer, Berlin (2014)
20. Neitzel, M., Mitschang, P., Breuer, U.: Handbuch Verbundwerkstoffe: Werkstoffe, Verarbeitung, Anwendung. Carl Hanser Verlag GmbH Co KG, Munich (2014)
21. Ehrenstein, G.W.: Faserverbund-Kunststoffe: Werkstoffe, Verarbeitung, Eigenschaften. Hanser Verlag, Munich (2006)
22. Scheydt, J.C.: Mechanismen der Korrosion bei ultrahochfestem Beton. KIT Scientific Publishing, Karlsruhe (2014)
23. Kreider, K.G., Patarini, V.M.: Thermal expansion of boron fiber-aluminum composites. Metall. Trans. **1**(12), 3431–3435 (1970)
24. Cichocki Jr., F., Thomason, J.: Thermoelastic anisotropy of a natural fiber. Compos. Sci. Technol. **62**(5), 669–678 (2002)
25. Akil, H., Omar, M., Mazuki, A., Safiee, S., Ishak, Z.M., Bakar, A.A.: Kenaf fiber reinforced composites: A review. Mater. Des. **32**(8–9), 4107–4121 (2011)
26. Hannemann, B.: Multifunctional metal-carbon-fiber composites for damage tolerant and electrically conductive lightweight structures IVW Publication series, vol. 128. Technische Universität Kaiserslautern, Institut für Verbundwerkstoffe GmbH (2017)
27. Suter-Kunststoffe AG: Basalt-, Fasern und Gewebe. Downloaded from: https://www.swiss-composite.ch/pdf/I-Basalt-Fasern-Gewebe.pdf. Downloaded on 12.07.2019 (2019)
28. Puck, A.: Zur Beanspruchung und Verformung von GFK-Mehrschichtenverbund-Bauelementen, vol. 57(4). Kunststoffe (1967)
29. Technische Universität Dresden, Institut für Luftfahrzeugtechnik: $eLamX^2$. Downloaded from: https://tu-dresden.de/ing/maschinenwesen/ilr/lft/elamx2/elamx. Downloaded on: 06.10.2018 (2018)
30. Fu, S., Lauke, B., Mai, Y.W.: Science and Engineering of Short Fiber Reinforced Polymer Composites. Elsevier Science, Amsterdam (2009)
31. Steffens, M.: Zur Substitution metallischer Fahrzeug-Strukturbauteile durch innovative Faser-Kunststoff-Verbund-Bauweisen. IVW Publication series Volume 14, Institut für Verbundwerkstoffe GmbH, TU Kaiserslautern (2000)

32. ULLMANN: Ullmann's Polymers and Plastics: Products and Processes. Wiley-VCH, Weinheim (2016)
33. Composites Institute: SPI/CI Introduction to Composites, 4th edn. Taylor & Francis, London (1998)
34. Usab, E.M., Usab, M.A.: Patentnummer EP0217841B1: Method and apparatus for the centrifugal casting of fiber reinforced plastic pIPD (1958)
35. HOBAS/Amiblu Germany GmbH: Schleuderverfahren für GFK-Rohre. Downloaded from: http://www.hobas.de/technologie/schleuderverfahren-cc.html. Downloaded on 11.09.2018 (2018)
36. Mitschang, P., Neitzel, M.: Handbuch Verbundwerkstoffe. Carl Hanser GmbH & Co. KG, Munich (2004)
37. Selden, P.H.: Glasfaserverstärkte Kunststoffe. Springer, Berlin/Heidelberg (2013)
38. Munro, M.: Review of manufacturing of fiber composite components by filament winding. Polym. Compos. 9(5), 352–359 (1988)
39. Shen, F.C.: A filament-wound structure technology overview. Mater. Chem. Phys. 42(2), 96–100 (1995)
40. Mack, J., Schledjewski, R.: Filament winding process in thermoplastics. In: Advani, S.G., Hsiao, K.-T. (eds.) Manufacturing Techniques for Polymer Matrix Composites (PMCs). Woodhead Publishing (2012)
41. Funck, R.: Entwicklung innovativer Fertigungstechniken zur Verarbeitung kontinuierlich faserverstärkter Thermoplaste im Wickelverfahren. VDI-Verlag, Düsseldorf (1996)
42. Anderson, J.V.: Automated Manipulation for the Lotus Filament Winding process. Thesis at Ira A. Fulton College of Engineering and Technology at Brigham Young University (2006)
43. Vargas Rojas, E., Chapelle, D., Perreux, D., Delobelle, B., Thiebaud, F.: Unified approach of filament winding applied to complex shape mandrels. Compos. Struct. 116(Supplement C), 805–813 (2014)
44. Industrieanzeiger (Article from 21.07.2014): Fachwerk aus der Wickelmaschine— Gewickelte Isogrid-Strukturen als völlig neue Leichtbau-Technologie. Downloaded from: https://industrieanzeiger.industrie.de/technik/entwicklung/fachwerk-aus-der-wickelmaschine/. Downloaded on 11.09.2018 (2014)
45. Kies, T.: 10 Grundregeln zur Konstruktion von Kunststoffprodukten. Carl Hanser Verlag GmbH & Company KG, Munich (2018)
46. Clegg, D.W., Collyer, A.A.: Mechanical Properties of Reinforced Thermoplastics. Springer, Netherlands (2012)
47. Vogt, D. (Hürth Nova-Institut für Ökologie und Innovation): Study: "Wood-Plastic-Composites"-Holz-Kunststoff-Verbundwerkstoffe. Downloaded from: www.nova-institut.de/pdf/06-01_WPC-Studie.pdf. Downloaded on 06.10.2018 (2006)
48. Baran, I.: Pultrusion: State-of-the-art Process Models. Smithers Information Limited, Akron (2015)
49. Meyer, R.: Handbook of Pultrusion Technology. Springer Science & Business Media, Berlin (2012)
50. Renkl, J.: Schneller geradeaus - und um die Kurve. K Magazin, vol. 01/2017 (2017)
51. Peters, S.T.: Handbook of Composites. Springer, US (2013)
52. R&G Faserverbundwerkstoffe GmbH: Handbuch Faserverbundwerkstoffe (2009)
53. IKV Aachen, Pressemitteilung, November 2015: Verfahrenskombination Pultrusion und Extrusion—Neue Werkzeugtechnik zur Funktionalisierung von pultrudierten FVK-Profilen. Downloaded from: http://www.ikv-aachen.de/neuigkeiten/detailseite-neuigkeiten/news/news/detail/verfahrenskombination-pultrusion-und-extrusion-neue-werkzeugtechnik-zur-funktionalisierung-von-pul/. Downloaded on 06.10.2018 (2015)
54. Britnell, D.J., Tucker, N., Smith, G.F., Wong, S.S.F.: Bent pultrusion—a method for the manufacture of pultrudate with controlled variation in curvature. J. Mater. Process. Technol. 138(1), 311–315 (2003)
55. thomas Technik + Innovation: Radius Pultrusion: Continuous Production of Curves. Kunststoffe International, vol. 11/2009 (2009)

56. Breuer, U.P.: Commercial Aircraft Composite Technology. Springer, Berlin (2016)
57. Eichenhofer, M., Wong, J., Ermanni, P.: Experimental investigation of processing parameters on porosity in continuous lattice fabrication. In: 21st International Conference on Composite Materials, Xi'an, China, 20.08.2017–25.08.2017 (2017)
58. Domm, M., Schlimbach, J.: Characterization of a novel additive manufacturing process for FRPC. SAMPE Europe Conference, Southhampton, United Kingdom, 11.09.2018–13.09.2018 (2018)
59. Beresheim, G.: Thermoplast-Tapelegen: Ganzheitliche Prozessanalyse und -entwicklung. IVW Publication series Volume 32, Institut für Verbundwerkstoffe GmbH, TU Kaiserslautern (2002)
60. Lukaszewicz, D.H.-J.A., Ward, C., Potter, K.D.: The engineering aspects of automated prepreg layup: History, present and future. Compos. B Eng. 43(3), 997–1009 (2012)
61. Brecher, C., Kermer-Meyer, A., Dubratz, M., Emonts, M.: Thermoplastische Organobleche für die Großserie. ATZextra 15(10), 28–32 (2010)
62. Khan, M.A., Mitschang, P., Schledjewski, R.: Tracing the void content development and identification of its effecting parameters during in situ consolidation of thermoplastic tape material. Polym. Polym. Compos. 18(1), 1–15 (2010)
63. Mack, J., Holschuh, R., Mitschang, P.: Qualitätsanalyse bei Bändchenhalbzeugen. Lightweight Des. 7(5), 48–53 (2014)
64. Holschuh, R., Becker, D., Mitschang, P.: Verfahrenskombination für mehr Wirtschaftlichkeit des FRP-Einsatzes im Automobilbau. Lightweight Des. 5(4), 14–19 (2012)
65. Campbell, F.C.: Manufacturing Processes for Advanced Composites. Elsevier Science, Amsterdam (2003)
66. Advani, S.G., Sozer, E.M.: Process Modeling in Composites Manufacturing, 2nd edn. CRC Press, Boca Raton (2010)
67. Malnati, P. (Composites World): Prepreg compression molding makes its commercial debut. Downloaded from: https://www.compositesworld.com/articles/prepreg-compression-molding-makes-its-commercial-debut. Downloaded on 19.09.2018 (2015)
68. Rimmel, O., May, D., Gemperlein, C., Mitschang, P.: Effects of fast prepreg pressing on laminate quality and mechanical properties. In: 21st International Conference on Composite Materials, Xi'an, PR China, 20.08.2017–25.08.2017 (2017)
69. Mennig, G., Stoeckhert, K.: Mold-Making Handbook. Carl Hanser Verlag GmbH & Company KG, Munich (2013)
70. Hildebrandt, K., Schulte-Hubbert, F., Mitschang, P.: Influence of textile parameters and laminate build-up on surface quality of thermoplastic fiber-reinforcced composites. In: 19th International Conference on Composite Materials, Montreal, Canada, 28.07. 2013–02.08.2013 (2013)
71. Hildebrandt, K.: Material- und prozessspezifische Einflüsse auf Oberflächeneigenschaften von endlosfaserverstärkten Thermoplasten. IVW Publication series Volume 116, Institut für Verbundwerkstoffe GmbH, TU Kaiserslautern (2015)
72. Breuer, U.: Beitrag zur Umformtechnik gewebeverstärkter Thermoplaste. VDI Verlag, Düsseldorf (1997)
73. Pudenz, K. (Automobil + Motoren): 800 g leichter: Lanxess zeigt Pkw-Sitzschale aus Hochleistungscomposite Tepex. Downloaded from https://www.springerprofessional.de/automobil—motoren/800-g-leichter-lanxess-zeigt-pkw-sitzschale-aus-hochleistungscom/6586498. Downloaded on 21.09.2018 (2013)
74. Edelmann, K.: CFK-Thermoplast-Fertigung für den A350 XWB. Lightweight Des. 5(2), 42–47 (2012)
75. Miaris, A., Edelmann, K., Sperling, S.: Thermoplastic matrix composites: Xtra complex, Xtra Quick, Xtra Efficient. In: 20th International Conference on Composite Materials, Copenhagen, Denmark, 19.07.2015–24.07.2015 (2015)
76. Fries, E., Renkl, J., Schmidhuber, S.: Mit vernetzter Kompetenz zum Hochleistungsbauteil. Kunststoffe 9, 52–56 (2011)

77. BMC Switzerland: Our best kept secret—the impec Advanced R&D Lab (Article from 29.10.2014). Downloaded from https://www.bmc-switzerland.com/int-en/experience/bmc-tempo/our_best_kept_secret_the_impec_advanced_rd_lab/. Downloaded on 21.09.2018 (2014)

78. Schnizer, M.: Anforderungen und Lösungsansätze für einen höheren Automatisierungsgrad in der CFK-Fertigung (Premium Aerotech GmbH). 2. Augsburger Produktionstechnik-Kolloquium, Augsburger, 15.05.2013 (2013)

79. Hildebrandt, K., Mack, J., Becker, D., Mitschang, P., Medina, L.: Potentiale neuer Matrixpolymere für die FRP-Bauteilfertigung. Lightweight Des. **7**(2), 14–21 (2013)

80. de Luzuriaga, A.R., Martin, R., Markaide, N., Rekondo, A., Cabañero, G., Rodríguez, J., Odriozola, I.: Epoxy resin with exchangeable disulfide crosslinks to obtain reprocessable, repairable and recyclable fiber reinforced thermoset composites. Mater. Horiz. **3**(3), 241–247 (2016)

81. Schommer, D., Duhovic, M., Hausmann, J.: Development of a solid mechanics based material model describing the behavior of SMC materials. In: 14th International Conference on Flow Processes in Composite Materials, Lulea, Sweden, 30.05.2018–01.06.2018 (2018)

82. Duhovic, M., Romanenko, V., Schommer, D., Hausmann, J.: Material characterization of high fiber volume content long fiber reinforced SMC materials. In: 14th International Conference on Flow Processes in Composite Materials, Lulea, Sweden, 30.05.2018–01.06.2018 (2018)

83. Schommer, D., Duhovic, M., Romanenko, V., Andrä, H., Steiner, K., Schneider, M., Hausmann, J. M.: Material characterization and compression molding simulation of CF-SMC materials in a press rheometry test. In: Key Engineering Materials (2019)

84. Braess, H.H., Seiffert, U.: Vieweg Handbuch Kraftfahrzeugtechnik. Vieweg + Teubner Verlag, Wiesbaden (2011)

85. Bonnet, M.: Kunststofftechnik: Grundlagen, Verarbeitung, Werkstoffauswahl und Fallbeispiele. Springer Fachmedien Wiesbaden, Wiesbaden (2016)

86. Schöpfer, J.: Spritzgussbauteile aus kurzfaserverstärkten Kunststoffen: Methoden der Charakterisierung und Modellierung zur nichtlinearen Simulation von statischen und crashrelevanten Lastfällen. Thesis, TU Kaiserslautern (2011)

87. Menges, G., Geisbüsch, P.: Die Glasfaserorientierung und ihr Einfluß auf die mechanischen Eigenschaften thermoplastischer Spritzgießteile—Eine Abschätzmethode. Colloid Polym. Sci. **260**(1), 73–81 (1982)

88. Bichler, M.: Prozessgrößen beim Spritzgießen: Analyse und Optimierung. Beuth Verlag GmbH (2012)

89. Witten, E., Mathes, V., Sauer, M., Kühnel, M.: Composites-Marktbericht 2018: Marktentwicklungen, Trends, Ausblicke und Herausforderungen. AVK—Industrievereinigung verstärkte Kunststoffe e.V. Carbon Composites e.V. (2018)

90. Witten, E., Kraus, T., Kühnel, M.: Compostes-Marktbericht 2015: Marktentwicklungen, Trends, Ausblicke und Herausforderungen. Ed.: AVK Industrievereinigung verstärkte Kunststoffe e.V. Carbon Composites e.V. (2015)

91. Lahr, R.: Partielles Thermoformen endlosfaserverstärkter Thermoplaste. IVW Publication series Volume 73, Institut für Verbundwerkstoffe GmbH, TU Kaiserslautern (2007)

92. Van Oosterom, S., Allen, T., Battley, M.A., Bickerton, S.: Evaluation of variety of vacuum assisted resin infusion processes. In: 21st International Conference on Composite Materials, Xi'an, China, 20.08.2017–25.08.2017 (2017)

93. Bader, M.G.: Selection of composite materials and manufacturing routes for cost-effective performance. Compos. A Appl. Sci. Manuf. **33**(7), 913–934 (2002)

94. Kaufmann, M., Berg, D. C., Greb, C., Cetin, M., Waeyenbergh, B., Jacobs, T.: Design for Manufacture for Liquid Composite Molding. TEXCOMP-11, Leuven, Belgium, 19.09.2013–20.09.2013 (2013)

95. Parton, H., Verpoest, I.: In situ polymerization of thermoplastic composites based on cyclic oligomers. Polym. Compos. **26**(1), 60–65 (2005)

96. Ishak, Z.M., Leong, Y., Steeg, M., Karger-Kocsis, J.: Mechanical properties of woven glass fabric reinforced in situ polymerized poly (butylene terephthalate) composites. Compos. Sci. Technol. **67**(3–4), 390–398 (2007)
97. Steeg, M.: Prozesstechnologie für Cyclic Butylene Terephthalate im Faser-Kunststoff-Verbund. IVW Publication series Volume 90, Institut für Verbundwerkstoffe GmbH, TU Kaiserslautern (2010)
98. Hopmann, C., Michaeli, W., Greif, H., Wolters, L.: Technologie der Kunststoffe. Carl Hanser Verlag GmbH & Co. KG, Munich (2015)
99. Deutsches Institut für Normung: DIN 7724—Polymere Werkstoffe (1993)
100. Gassmann, O., Kobe, C.: Management von Innovation und Risiko: Quantensprünge in der Entwicklung erfolgreich managen. Springer Science & Business Media, Berlin (2006)

Chapter 5
Phase 3: Elaboration

Abstract This section describes the third phase of the proposed approach to integrated product development with FRPs. In this phase the drafts are elaborated, which includes the final selection of all materials and semi-finished products to be used. Furthermore, the design is elaborated while taking into account production-, joining-, repair- and recycling-related aspects. Finally, this section deals with the elaboration of the process concept, i.e. the selection of production systems and the definition of quality assurance measures.

5.1 Overview

Figure 5.1 shows the procedure of phase 3, for the elaboration of the drafts. The individual steps are described in detail in the following.

5.2 Elaboration of the Material Concept

The first step of the draft elaboration is the completion of the material selection. In the concept/draft phase, the fiber material was defined and it was determined whether a thermoset or thermoplastic polymer is to be used. To complete the material selection, the semi-finished products and the specific polymer type must be defined. After this, the material properties required for component design must be determined.

5.2.1 Selection of Semi-finished Products

For the production of FRPs, various semi-finished products can come to use. The selection of semi-finished products represents both risk and opportunity. A bad

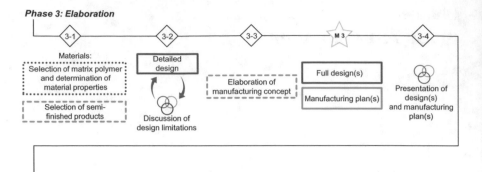

Fig. 5.1 Overview of phase 3—elaboration

choice induces a high risk of a lack of process efficiency. On the other hand, a targeted selection, based on sound scientific findings, offers a great opportunity to increase process efficiency. For this reason, the competences of the different involved departments should be bundled for the selection of the semi-finished products. The team member from the manufacturing department should lead this selection process, as the effect of the manufacturing process must be properly considered.

In general, three basic categories of semi-finished products can be distinguished for the manufacture of FRPs:

1. **Dry and bindered semi-finished products**, which are exclusively made of fibers and auxiliary materials for shape fixation and/or bonding.
2. **Matrix materials**, i.e. matrix polymers, possibly chemically modified or filled.
3. **Pre-impregnated semi-finished products** in which fibers and matrix polymer are combined.

As shown in Fig. 5.2, these categories are subdivided into various sub-categories and groups. The illustration is not necessarily complete, since—as so often when dealing with FRPs—an increasing diversification can be observed. The intention behind the figure is to give an overview of the most important semi-finished products.

The selections concerning the structure of the fiber reinforcement, the matrix polymer class (see Sect. 4.6) and the manufacturing process already strongly pre-define the suitable types of semi-finished products. Table 5.1 shows possibly useable semi-finished products for the different processes. The use of semi-finished products that are not listed in this table is not necessarily technically impossible, but for various reasons—mostly lack of cost-efficiency—unusual. If only a group or subgroup is listed, all semi-finished products of this group can be used.

With the help of this table, the semi-finished product can already be preselected. For further specification, the following aspects should be particularly considered:

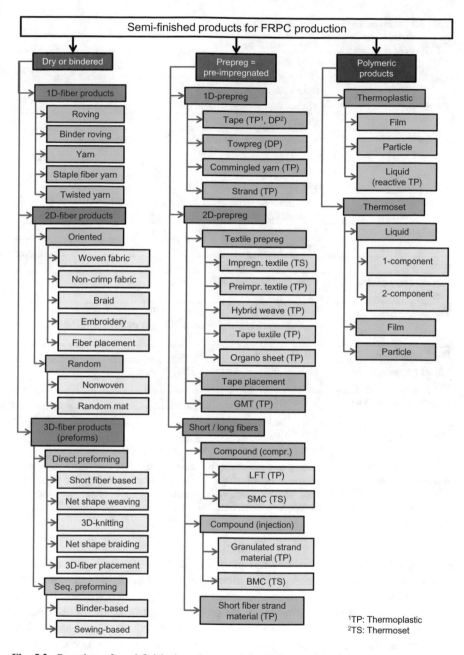

Fig. 5.2 Overview of semi-finished products used for FRP manufacturing

Table 5.1 Semi-finished products used within the different processes

Process	Semi-finished products
Fiber spraying	(Rovings)+(resin system)
Centrifugal casting	(Rovings, 2D non-wovens, woven fabric, non-crimp fabric) + resin system
Winding	TS: rovings + resin system or TS-tapes or towpreg
	TP: TP-tapes
Pultrusion	TS: (rovings, limited usage of 2D non-wovens and woven fabric) + resin system
	TP: commingled yarns
Extrusion	TP: granulated strand material (TP)
3D printing (cont. reinforcement)	TP: strand material, short or continuous reinforced, roving + TP-melt (from particles)
Tape laying	TS: TS-tapes (requires curing cycle)
	TP: TP-tapes
Autoclave	TS: TS-tape or 2D prepreg
Prepreg compression molding	Impregnated textile (TS), TS-tapes
Thermoforming	Isothermal[a]: organo sheet, tape textile Variothermal[b]: hybrid textile, 2D fiber product + (TP-films, TP-particles)
Hand lamination	2D fiber product + resin system
Liquid composite molding	(2D/3D fiber product + resin system)
Compression molding	TS: SMC
	TP: LFT, GMT
Injection molding	TS: BMC
	TP: strongly granulated strand material (TP)

[a]Here the temperature of the forming tool is virtually constant and below the melting temperature, so that rapid solidification is achieved. However, this means that impregnation is not possible
[b]The temperature of the forming tool is variable. Impregnation can take place by temporary tempering over the melting temperature. Afterward the tool is cooled down again to achieve solidification

- **Design issues**: Semi-finished products with similar appearance can still show remarkable differences concerning the performance of the final component. For example, due to fiber undulation 3D-knitted fabrics provide considerably worse mechanical properties compared to 3D preforms manufactured by fiber placement (straight fibers). On the other hand, unexpected semi-finished product characteristics can be challenging for product designers. For example, fabrics that are symmetric according to the data sheet often provide mechanical anisotropy, because of the manufacturing process, where warp and weft threads are undulated to varying degrees. Therefore, it is important to consider the semi-finished products with this background in mind.

- **Manufacturing-related issues**: Equivalently, the processing properties can vary quite considerably, even for otherwise substitutable semi-finished products. Compared to woven or non-crimp fabrics, fiber structures manufactured by fiber placement lead to remarkably longer cycle times when the process conditions are the same [1]. At the same time, the waste can be greatly reduced [2]. Such points must be taken into account at this point so that the economic and technical meaningfulness of the manufacturing concept is not compromised.
- **Economic issues**: For many semi-finished products, certain standard material combinations have been developed, which are common and therefore available in relatively large quantities and comparatively inexpensive. If an alternative material combination is required, it must be specially produced, which can quickly lead to a significant price increase. In addition, from an economic point of view, possibilities for recycling, minimization of waste, storage costs and much more must be considered.

In the following section, some important semi-finished products will be presented. According to the intention of the IPD, the most important respective design, manufacturing and economic issues are described, in order to support the selection process.

5.2.1.1   Dry Semi-finished Products

1D Semi-finished Fiber Products (Unidirectional)

1D semi-finished fiber products form the first stage of semi-finished products, starting from the single filament. They bundle numerous fibers (usually several thousand). One can distinguish:

- **Roving**: In rovings, several thousand fibers are unidirectionally bundled. Depending on the design of the roving, the fibers can be more or less spread. The sizing applied to the fibers leads to a light cohesion.
- **Bindered roving**: A spread roving, which is held together by a polymeric binder (usually in powder form).
- **Multi-end roving**: Here several rovings are jointly placed on a spool. There is no cohesion between the individual rovings.
- **Yarn**: Describes a roving twisted along the longitudinal axis. Partly also referred to as "twisted" or "protective twist" (in textile terminology), because it improves cohesion.
- **Staple fiber yarn**: 1D semi-finished product made of staple fibers (short/long fibers). Fibers that are only available as staple fibers (natural fibers) can be processed to a staple fiber yarn.
- **Twisted yarn**: Two or more rovings twisted together.

Table 5.2 To be noted for 1D fiber products

Semi-finished product	To be noted concerning…		
	…design	…manufacturing	…economics
Roving, In general for 1D fiber	• Matrix compatibility of fiber sizing critical for mechanical performance of composite • Mechanical properties remarkably smaller than theoretically derived from single fiber values, due to imperfect orientation etc.	• Output per time tends to increase with increasing linear mass density[a] • Increasing linear mass density tends to be increasingly challenging in production	• Costs tend to decrease with increasing linear mass density
Binder roving	• Binder can affect matrix properties remarkably	• Roving stability depends on binder type, amount, distribution and activation process • Binder can affect matrix processing behavior	• Only standard combinations broadly available → other combinations have to be individually produced
Multi-end roving	• Width variation	• Gap formation during spreading	
Yarn	• Property reduction due to twists [3]	• Increased stability but limited spreading possibilities	
Staple fiber yarn	• Performance reduction through limited fiber length and orientation	• Staple fibers can slide from each other→ plastic deformation possible [4]	• Potential option for recycling of relatively cheap CF-scrap
Twisted yarn	• Property reduction due to twists [3]	• Increased stability but limited spreading possibilities	

[a]Linear mass density is the weight per unit of length, typically given in the unit tex, where 1 tex equals 1 g/km

Table 5.2 lists issues to be noted concerning 1D semi-finished fiber products.

2D Semi-finished Fiber Products (Flat)

Here, a first categorization can be made with regard to the fiber orientation. "Oriented" semi-finished fiber products are manufactured from 1D semi-finished fiber products. The rovings are oriented, and their length in the semi-finished product is virtually endless. In "random" 2D semi-finished fiber products, the fibers are not specifically oriented. Within these two categories, different types can be distinguished:

- **Oriented 2D semi-finished fiber products**

  - **Woven Fabrics**: Here, two rovings are interwoven at an angle of 90° whereby different weaves (structure of the interlacing) are possible. In addition, rovings made of different materials can be combined (e.g. CF + AF). Note that 3D fabrics must be distinguished. In 3D fabrics, additional rovings oriented in thickness direction are interwoven, resulting in multilayer structures.
  - **Non-crimp fabric**: Usually one to four stacked layers (more are also possible) of unidirectional oriented rovings are fixed by a relatively thin secondary yarn (usually made of polyester). This creates a multi-axial stacking, in which fibers of different, but specific orientation are combined.
  - **Braiding**: Flat structures of interlaced rovings, where the braiding angles can be varied and different rovings can be combined. For the production, initially a sleeve is braided using a braided ring and then slit.
  - **Embroidery**: In a sewing process, a roving is stitched on a base material (e.g. a very light GF fabric). Thereby, the orientation can be varied freely, which also allows the construction of locally load-adapted preforms.
  - **Fiber Placement**: Bindered rovings placed on a tool via a tape laying process (see Sect. 4.6.2.7). The locally melted binder fixes the rovings to each other. The discontinuous, direction and position variable placement allows the production of locally load-adapted preforms.

- **Random 2D semi-finished fiber products**

  - **Non-woven:** Processed out of a fiber bunch (e.g. natural fibers), in which finite fibers of different lengths are present. Because of the processing, shortening and slight alignment takes place. The fixation is achieved by polymeric binder or by needling.
  - **Random mat:** Here, endless fibers (e.g. glass fibers) are randomly arranged. They can be distinguished between endless fiber mats (continuous fibers) and chopped fiber mats (short/long fibers). If necessary they are fixed chemically or by needling.

Table 5.3 lists issues to be noted concerning 2D semi-finished fiber products.

3D Semi-finished Fiber Products

3D semi-finished fiber products, also known as preforms, due to their near-net shape, are produced in a separate process step, the so-called preforming. They are produced either from 2D semi-finished fiber products (sequential preforming) or directly from 1D semi-finished fiber products (direct preforming) [5].

- **Direct preforms**

  - Short fiber-based: Various processes allow to randomly place short fibers on a tool with the contour of the later component. The fibers are, e.g., sprayed on. Alternatively, the fibers are put in a water bath together with the tool, which is then perforated. By sucking the water through the perforation, the

Table 5.3 To be noted for 2D fiber products

Semi-finished product	To be noted concerning...		
	...design	...manufacturing	...economics
Woven fabric	• Reduction in mechanical properties due to roving undulation (depending on type of weave) • Depending on type of weave, symmetry can only be reached with an even number of layers • Hybrids with multiple fiber materials possible • 3D-woven fabrics: reduced risk of delamination, improved out-of-plane properties, reduced in-plane properties	• Stability and drapeability depend on type of weave (trade-off!) • Anisotropic impregnation behavior due to roving crimp variations • Increasing areal weight reduces effort for lay-up but makes draping more difficult • 3D-woven fabrics: Very high areal weights can be reached but drapeability is strongly limited as single layers cannot slide past each other	• Relatively high material input rate[a] • Classic "carbon look" (twill weave 2/2) • Relatively cheap due to high production volumes and high manufacturing efficiency
Non-crimp fabric	• Undulation of fibers often reduced compared to woven fabrics • Fiber material combinations possible • Fiber angles are commonly between 20° and 90°, related to production direction (0°) as well as directly in production direction	• Complete lay-up in a single layer → high accuracy • Low heat stability of secondary yarn • Stability and drapeability depend on stitching type (trade-off!) • Increasing areal weight reduces effort for lay-up but makes draping more difficult • Flow channels induced by secondary yarn improve impregnation behavior	• Relatively high material input rate • Relatively cheap due to high production volumes and high manufacturing efficiency
Braidings	• Reduction in mechanical properties due to roving undulation • Fiber material combinations possible	• For hollow parts, changes in diameter change the fiber angles	• Relatively high material input rate when used as 2D fiber product

(continued)

Table 5.3 (continued)

Semi-finished product	To be noted concerning…		
	…design	…manufacturing	…economics
Embroidery	• Locally load-adapted fiber positioning possible • No fiber undulation	• High output can only be reached by extensive parallelization • Limited heat resistance of stitching yarn	• Relatively slow process → high output only through extensive parallelization • Relatively low material input rate • Special preform designs required that possibly have to be procured from service providers
Fiber placement	• Locally load-adapted fiber positioning possible • No fiber undulation	• Bound only by binder → limited heat resistance • Impregnation behavior very challenging • Draping behavior challenging • Binder can affect processing behavior of matrix	• Relatively slow process → high output only through extensive parallelization • Relatively low material input rate • Special preform designs required that possibly have to be procured from service providers
Non-woven/ random mat	• Quasi-isotropic to slightly anisotropic behavior • Reduction in mechanical properties due to roving undulation • Good damping properties	• Plastic deformation can lead to local thinning • Relatively "fluffy," hence compaction resistance is typically quite high	• Possible option for recovery of CF-scrap (but downcycling) • Classic type of fiber product for NF

[a]The material input rate is the ratio of the amount of material used for the production of a part and the amount of material that actually ends up in the final part. It therefore quantifies the fraction of waste during manufacturing (100% would correspond to zero waste)

fibers are pulled onto the tool. The cohesion of the preform is achieved by polymeric binders.

– Contour weaving: Advanced weaving technologies allow weaving of certain final contours, for example, profiles, directly out of rovings.

– Contour braiding: Rovings are braided with a braiding ring, i.e. a sleeve is created. This also allows for changes in cross section and curvature. For this, the braiding core must be accordingly shaped and guided, e.g., by an industrial robot.

- 3D-knitting: Here the rovings are interlaced in loops. Just like in classic knitting for clothing, complex geometries can be produced, as the lacings provide for very high drapeability.
- 3D fiber placement: Here, binder rovings are processed in a tape laying process (see Sect. 4.6.2.7), whereby the locally melted binder is used for fixation. The discontinuous, direction and position variable placement allows the production of locally load-adapted preforms. By depositing on a correspondingly formed tool (shell-shaped), 3D preforms can be produced directly out of rovings.

- **Sequential preforming:** In this process, 2D semi-finished fiber products are brought into a near-net shape. For this, they are stacked to the required layer structure and cut in shape. This is followed by shaping and fixing. Finally, where appropriate, the sub-preforms are merged to the final preform. For fixation and sub-preform joining, there are two basic alternatives.

  - **Binder technology**: Fixation is realized by a polymeric binder, which is usually present in the form of particles or webs. The binder is applied to the fabric either directly during the production of the 2D semi-finished fiber product or later. Before shaping, the binder is melted and then, when the desired shape is reached, solidified again.
  - **Sewing technology**: Fixation of individual layers and sub-preforms is realized by sewing.

Table 5.4 lists issues to be noted concerning 3D semi-finished fiber products.

5.2.1.2   (Pre-)Impregnated Semi-finished Products (Prepregs)

In (pre-)impregnated semi-finished products, commonly referred to as prepregs, reinforcing fibers and matrix polymer are already combined. Thermoplastic and thermoset prepregs differ in one important point: meltability. Thermoplastic (TP) prepregs contain matrix polymer in the same chemical form, as it will later also be present in the component. For further processing, the TP prepreg is melted again. In thermoset (TS) prepregs all components of the matrix polymer (e.g. resin and hardener) are present, but mostly uncured, since re-melting is not possible. The cross-linking reaction is not completely prevented, but rather slowed down by appropriate storage conditions (low temperature). Accordingly, shelf life of TS prepregs is limited, while TP prepregs can be stored virtually infinitively. Despite the differences, the available prepregs are quite similar in appearance, which is why in the following a subdivision according to the structure of the fiber reinforcement (1D, 2D and random oriented short and long fibers) and according to the matrix polymer is given.

1D Prepregs

1D prepregs are processed out of rovings. The unidirectional reinforcement structure is preserved, and the roving is (pre-)impregnated with a matrix polymer.

Table 5.4 To be noted for 3D fiber products

Semi-finished product	To be noted concerning…		
	…design	…manufacturing	…economics
In general for 3D fiber products (preforms)	• The preforming technology can strongly affect the mechanical performance	• The preforming technology can strongly affect the processing behavior (e.g. impregnation behavior) • High relevance of dimensional accuracy: → too small preforms can lead to displacement in further processes and hence deviations concerning fiber orientation → too large preforms can cause dry spots, porosity and fiber disorientation	• For relatively complex preforms a decision concerning preforming depth (number of sub-preforms) has to be made: Costs for preforming effort and the part cost reduction have to be evaluated
Short fiber-based	• Limited fiber length and orientation lead to reduced mechanical properties • Quasi-isotropic behavior	• Tool perforation (if necessary) clogs fast • Challenging reproducibility when based on manual work	
Net shape weaving	• Reduction of mechanical properties due to fiber undulation • Reduced delamination behavior • Not all fiber orientations manufacturable • Almost entirely for profiles • Material combinations possible	• Often limited to research level	• Relatively slow production • Relatively low material input rate
Net shape braiding	• Reduction of mechanical properties due to fiber undulation • Not all fiber orientations manufacturable	• Core required (possibly has to stay within the part)	• Relatively low material input rate

(continued)

Table 5.4 (continued)

Semi-finished product	To be noted concerning…		
	…design	…manufacturing	…economics
	• Almost entirely for hollow profiles • Variations on cross section possible • Material combinations possible		
3D-knitting	• Strong reduction in mechanical properties due to fiber undulation • Very high geometric flexibility	• Net shape manufacturing possible • Cutting problematic (ladder!) • Extreme drapeability • No fringe-out when not cut (closed edges) • Small radii critical for CF fibers	• Relatively low material input rate
3D fiber placement	• Locally load-adapted fiber orientation possible • No fiber undulation • Binder can affect mechanical properties of matrix	• Bound only by binder → limited heat resistance • Impregnation behavior very challenging • Binder can affect processing behavior of matrix	• Relatively slow process → high output only through extensive parallelization • Relatively low material input rate • Special preform designs required that possibly have to be procured from service providers
Binder-based	• Binder can affect mechanical properties of matrix • Fiber orientation defined by forming → draping simulation!	• Preform stability depends on binder type, amount, distribution and activation process • Binder can affect processing behavior of matrix • Danger of wrinkling due to forming • Edge trimming of forming required	• Suitable for large bath production, due to high automation potential • Bindering can be done already on 2D fiber products • Press required for forming (relatively high investment) • Relatively high material input rate
Sewing-based	• Sewing threads affect mechanical performance → chance (e.g. reduce risk of delamination) and risk (e.g.	• Sewing threads affect impregnation and draping behavior → chance (e.g. increased permeability, due to	• Sewing, as an additional process step, is often not cost-efficient → multi-functionality is the key to

(continued)

Table 5.4 (continued)

Semi-finished product	To be noted concerning…		
	…design	…manufacturing	…economics
	possible crack initiator) • Functional integration (e.g. fixing of load introduction elements)	flow channels) and risk (e.g. unwanted flow front distortion) • Pre-compaction through sewing can simplify tool loading	cost-efficiency (e.g. sewing for shape fixation and improved mechanical performance)

- **Tape**: flat, spread roving embedded in a thermoset or thermoplastic matrix.
- **Towpreg**: roving pre-impregnated with a thermoset resin system.
- **Strand material**: roving pre-impregnated with a thermoplastic polymer.
- **Hybrid roving (commingled yarn)**: roving in which reinforcing fibers and polymer fibers (matrix polymer) are combined.

Table 5.5 lists issues to be noted concerning 1D prepregs.

2D Prepregs

2D prepregs are manufactured either from 1D prepregs or by impregnation or hybridization of a 2D semi-finished fiber product with a semi-finished matrix product.

- **Textile prepregs**

  - **Impregnated textile (TS)**: textile impregnated with a thermoset resin system (mostly woven or non-crimp fabric). Due to its popularity, the term "prepreg" is often used synonymous for this type of semi-finished product.
  - **Pre-impregnated textile (TP)**: semi-finished textile product pre-impregnated with a thermoplastic matrix polymer. The matrix polymer can be applied as powder, dissolved in solvent or directly as a melt.
  - **Organo sheet**: textile embedded in a thermoplastic matrix (almost exclusively woven fabric, since this ensures cohesion even in the melted state), which is processed out of a pre-impregnated textile (TP) or a hybrid textile. Organo sheets are (almost) fully impregnated and consolidated.
  - **Hybrid textile**: textile (fabric) made from hybrid yarns.
  - **Tape textile**: Woven fabric made of thermoplastic tapes.

- **Tape placement:** Semi-finished products made of thermoplastic tapes via thermoplastic tape laying or an alternative joining method (e.g. spot welding). Depending on the process, a fully consolidated semi-finished product or a loosely joined semi-finished product is achieved.
- **Glass mat-reinforced thermoplastic (GMT):** 2D fiber semi-finished product with random fiber orientation embedded in a thermoplastic matrix.

Table 5.5 To be noted for 1D prepregs

Semi-finished product	To be noted concerning…		
	…design	…manufacturing	…economics
Tape	• Locally load-adapted design possible • Suitable for high-performance parts	• TS: Manual and automated processing possible; placement on tool supported by temperature-dependent tack • TP: Virtually only automated processing; only limited post-impregnation ($\to$ tape quality is crucial)	• TS: Limited shelf life, storing facilities required • Use standard material combinations if possible!
Towpreg	• Virtually only used for winding		• Limited shelf life, storing facilities required • Use standard material combinations if possible!
Hybrid roving		• Short flow paths but post-impregnation is mandatory	• Use standard material combinations if possible!
Strand material		• Only limited post-impregnation in 3D printing ($\to$ tape quality is crucial)	• Standard material combinations are relatively cheap

Table 5.6 lists issues to be noted concerning 2D prepregs.

Randomly Oriented Short/Long Fibers (Compression/Injection Molding Compounds)

Compounds are required for compression and injection molding processes.

- **Long fiber-reinforced thermoplastics (LFT)**: granulated thermoplastic strand material (fiber length usually between 10 and 25 mm).
- **Injection-moldable, highly granulated strand material**: highly granulated strand material in order to achieve the flowability required for injection molding (fiber length usually below 1 mm).
- **Sheet molding compound (SMC)**: Compound made from cut fibers (length usually between 25 and 50 mm), a thermoset matrix and fillers.
- **Bulk molding compound (BMC)**: Similar to SMC, with shorter fiber length and therefore partially processable in injection molding (fiber length usually less than 12 mm).

Table 5.7 lists issues to be noted concerning compression and injection molding compounds.

Table 5.6 To be noted for 2D prepregs

Semi-finished product	To be noted concerning…		
	…design	…manufacturing	…economics
Impregnated textile (TS)	• Equivalent to dry textiles (Sect. 5.2.1.1)	• Placement on tool supported by temperature-dependent tack • Equivalent to dry textiles (Sect. 5.2.1.1)	• Relatively high material input rate • Use standard material combinations if possible!
Pre-impregnated textile (TP)	• Equivalent to dry textiles (Sect. 5.2.1.1)	• Type and degree of pre-impregnation strongly affect the required cycle time for full impregnation and consolidation • Direct processing requires variothermal tool temperature • Equivalent to dry textiles (Sect. 5.2.1.1)	• Continuous processing to organo sheets (suitable for large series) is possible
Organo sheet	• Equivalent to dry textiles (Sect. 5.2.1.1)	• Risk of wrinkling when formed • Surface quality depends on temperature control • Equivalent to dry textiles (Sect. 5.2.1.1)	• Relatively high material input rate • Use standard material combinations if possible!
Hybrid textile	• Equivalent to dry textiles (Sect. 5.2.1.1)	• High drapeability • Short flow paths but post-impregnation is mandatory • Direct processing requires variothermal tool temperature • Equivalent to dry textiles (Sect. 5.2.1.1)	• Relatively high material input rate • Use standard material combinations if possible!
Tape textile		• Limited drapeability • Structurally stable even when matrix is molten	• Relatively high material input rate
Tape placement	• Locally load-adapted design possible • Virtually undulation free	• Only bound by binder → Deformation at elevated temperatures possible	• Relatively low material input rate
GMT	• Typically, quasi-isotropic • Limited fiber length and orientation lead to reduced mechanical properties	• Limited flowability	• Widely replaced by SMC/LFT

Table 5.7 To be noted for compounds

Semi-finished product	To be noted concerning…		
	…design	…manufacturing	…economics
LFT/ granulated strand material	• Quasi-isotropic behavior, but local anisotropy possible due to flow-induced fiber orientation • Limited fiber length and orientation lead to reduced mechanical properties • Flowability allows for rib structures	• Flowability strongly affected by fiber length	• Standard material combinations are relatively cheap
SMC/BMC	• High surface quality possible • Quasi-isotropic behavior, but local anisotropy possible due to flow-induced fiber orientation • Good electric insulation • Fillers critical for part properties • Limited fiber length and orientation lead to reduced mechanical properties • Flowability allows for rib structures	• Maturing time required for thickening • Limited shelf life • Flowability strongly affected by fiber length • Processing behavior strongly affected by filler	• (Cooled) storing facilities required • SMC: Recycling option for CF-scrap

5.2.1.3   Semi-finished Polymer Products

Semi-finished polymer products are either used to manufacture prepregs or FRP components directly. A distinction between thermoplastic and thermoset matrix polymer semi-finished products is reasonable.

Thermoplastic

Thermoplastic semi-finished polymer products are re-meltable, which is exploited during FRP production. Due to the high melt viscosity, a good distribution on the semi-finished fiber product is important in order to keep the impregnation flow paths short.

- **Films**: The processing of thermoplastics into films allows an alternating layer structure with semi-finished fiber products, so that the flow paths during impregnation are limited to the individual layer thickness of the semi-finished fiber products.
- **Particles**: By granulation or pulverization, a good distribution can be achieved. The particle size depends on the application. For the production of a powder prepreg as a preliminary stage of an organo sheet, or when using binder technology for preforming relatively small binder particles are used. For processing in an extruder, for example, in the production of LFT, coarser granules are used.

- **Liquid (in situ polymerizing)**: In situ polymerization means that the thermoplastic polymer is only formed out of precursors (oligomers) in the FRP production process (e.g. a liquid impregnation process). This is possible with special thermoplastic systems. For example, polyamide 6 is produced from caprolactam [6].

Thermoset

Thermoset matrix systems cross-link (cure) during the course of FRP production. There are two possibilities of processing. Either the components required to initiate the cross-linking reaction (e.g. resin and hardener) are brought together directly in the component production process, or the components are already combined and the cross-linking reaction is slowed down by suitable storing conditions (especially low temperature). In both cases, often a temperature increase is used to initiate or accelerate the cross-linking reaction.

- **Liquid**: Two-component systems are particularly common. Usually at least one of the components is present in liquid form, often all of them. Mixing in a specific ratio produces the reactive mixture, which cross-links to form the thermoset. Some resin systems are also available as so-called 1-component systems, which are ready-mixed. This reduces the risk of a faulty mixture ratio. A popular example is given by HexFlow® RTM6 from Hexion, which is widely used in aviation. Partially these systems are solid at room temperature, and for processing, they must first be liquefied by heating.
- **Film**: Ready-mixed thermoset resin system in film form can be alternately stacked with 2D fiber semi-finished products and then pressed. In this way, short flow paths are achieved. The films are widely uncured and liquefy during the manufacturing process, thus impregnating the 2D semi-finished fiber products.
- **Particles**: Some thermoset resin systems are also available as particles. They are mainly used as binders in binder forming technology for preforming. Compared to thermoplastic binders, they offer improved matrix compatibility and, through cross-linking, better preform stability.

The basic issues to be noted regarding thermoplastic and thermoset materials have already been explained in Sect. 4.7. Table 5.8 lists issues to be noted concerning semi-finished matrix polymer products.

5.2.1.4 Characterization of the Processing Behavior of Semi-finished Products

Characterizing the processing behavior of semi-finished products is crucial for a proper selection but also for the later design. On the one hand, characterization makes it possible to test the suitability for the intended application, for example, when it comes to the draping behavior of a textile. On the other hand, values relevant for the later cycle time estimation can be determined. However, the variety of semi-finished products is enormous and the behavior is very complex due to the

Table 5.8 To be noted for polymers as semi-finished products

Semi-finished product	To be noted concerning…		
	…design	…manufacturing	…economics
TP-particles	• Particles can cause fiber deformations and corresponding reduction in mechanical performance		• Use standard material combinations if possible!
TP-films		• Film thickness must be adapted to applied 2D fiber products • Material thins when stretched (if this is possible) • Stretched films can shrink when heated	
TP-liquids (in situ polymerizing)		• New and complex facilities required • Relatively sensitive processes	• Cheaper than polymerized material • Material responsibility is with FRP part manufacturer
TS-liquid		• Exothermal reaction! • Two-component: accurate mixing ratio must be reached	• One-component: High transport costs as dangerous goods with required cooling
TS-film		• Film thickness must be adapted to applied 2D fiber products • Liquidation when heated	
TS-particles	• Degree of pre-curing determines cross-linking with matrix polymer	• Degree of pre-curing determines preform stability	

heterogeneous structure based on thousands and thousands of filaments. In many areas, there are no uniform standards for the characterization of material properties and the standards developed for other materials are often not transferable. In addition, the corresponding test rigs are frequently subject of research themselves. The acquisition of reliable data is therefore challenging. In order to give an overview, some semi-finished product processing characteristics and corresponding test methods are briefly presented below.

Impregnation Behavior of Fiber Structures

Manufacturing FRPs requires to bring fibers and matrix polymer together, when producing either semi-finished products or the final component. For continuous fiber reinforcement, e.g. by a textile, this means that the fiber reinforcement must be impregnated with the matrix. This corresponds to the flow of a fluid in a porous media, which can be described by the law of Darcy (Eq. 5.1). The law of Darcy was empirically determined by Henry Darcy in 1865 [7]. It allows calculation of the volume flow rate Q in a porous media, resulting from a pressure drop Δp over a flown-through length ΔL and given the fluid viscosity η, the flown-through cross-sectional area A and the permeability K of the porous media.

$$Q = -\frac{K \cdot \Delta p \cdot A}{\eta \cdot \Delta L} \qquad (5.1)$$

The impregnation behavior of the porous medium, in this case the fiber structure, is therefore quantified by the permeability, which thus describes the conductivity for fluid flow. The permeability is direction-dependent and hence for the three-dimensional space defined by a second-degree tensor. Considering textile symmetry conditions and aligning the tensor to the main flow directions, the tensor can be reduced to three permeability values, which must be known so that flow in all spatial directions can be described [8]. These three values are the highest and lowest in-plane permeability and the through-thickness permeability. Several technologies were developed to determine the permeability. Yet, there is no uniform standard. However, there are ongoing efforts to unify measurement approaches (see [9–17]). Permeability is typically measured by generating a fluid flow in a sample of the fiber structure of interest using a test fluid with known viscosity (often oils). One option is to generate a continuous flow through a sample with defined cross section and measure the volume flow rate resulting from a certain pressure drop over the sample length (saturated measurement). Another option is to record the flow front geometry and velocity for an injection under defined conditions (unsaturated measurement). Attention must be paid to the validity criterions of the law of Darcy (see Sect. 5.4.2.1). Figure 5.3 shows examples of measuring systems used for unsaturated through-thickness permeability and in-plane permeability characterization. For the latter, a so-called "radial injection" approach is used, where the flow ellipse resulting from a central point injection is recorded. This method allows measurement of the highest and lowest in-plane permeability and their orientation in one step.

The permeability of a fiber structure is very much dependent on the available pore space and thus the fiber volume content. Accordingly, on the one hand, it is necessary to determine the permeability values as a function of the fiber volume content and on the other hand, the fiber volume contents actually present in the process need to be known. In many processes, the fiber volume content is determined by the cavity in which the preform is compacted. Others have variable fiber volume contents. The vacuum infusion processes are interesting in this context, because here the preform is placed under a vacuum bag and compacted through the

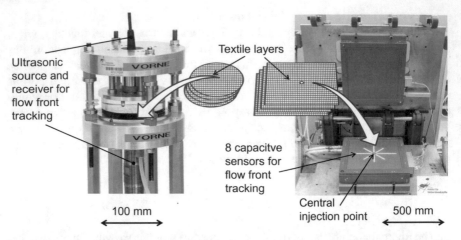

Fig. 5.3 Measurement systems for the characterization of the unsaturated through-thickness (left) and in-plane (right) permeability. Printed with permission of Leibniz-Institut für Verbundwerkstoffe GmbH

atmospheric pressure. In order to predict the fiber volume content under these conditions, the textile compaction behavior must be known [18].

Forming Behavior of Fiber Structures and Continuously Reinforced Pre-impregnated Semi-finished Products

Especially for geometrically complex components, with curved and double-curved surfaces, edges, corners, etc., it is important to take into account the forming behavior of fiber and fiber-matrix semi-finished products, when selecting them. There are a number of parameters, which quantify this behavior and can be useful in the selection process.

Draping mechanisms are one aspect of the forming behavior. Figure 5.4 shows the most important draping mechanisms, which can occur in the individual layers of a textile. These occur when handling a textile and, above all, when bringing them into a three-dimensional shape, whether intentionally or unintentionally. A good example of this is in-plane shearing, which is one of the mechanisms that occur during thermoforming [19].

Shearing refers to an angular change between the different fiber orientations in a textile. Only by shearing, is it possible to form three-dimensional shapes without wrinkles. This becomes clear when wrapping a bottle of wine in wrapping paper. The bottleneck can only be formed by folding, as the wrapping paper does not have the ability to shear. Textiles can differ very much in terms of drapeability, especially shearability, depending on their structure and the applied fiber sizing. It is important to consider this. If a high drapeability is not required, the textile should not provide it, as this increases the risk of unwanted shearing and thus fiber disorientation. On the other hand, too little drapeability could lead to wrinkling during production. There are various test methods available for determination of the forces required for shearing and the maximum shearing angles that can be achieved without wrinkling.

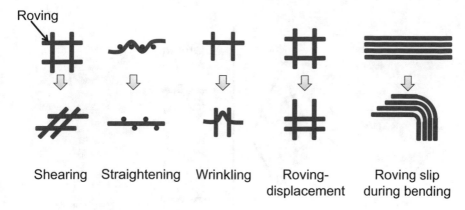

Roving

Shearing   Straightening   Wrinkling   Roving-displacement   Roving slip during bending

Fig. 5.4 Draping mechanisms in textiles

The "picture frame" and the "bias extension" tests [20] are two common examples, although neither of them have been standardized. In addition, there are more specific test methods, such as the "Drapetest" system from Textechno [21] for woven and non-crimp fabrics, which, in addition to drapeability under defined tension conditions (retention forces), also allows the testing of the other above-mentioned effects (e.g. gap formation due to roving displacement). Furthermore, DIN SPEC 8100 is a first attempt at standardization of such tests, while an ISO standardization is also in preparation. Figure 5.5 shows visualizations of these procedures.

Generally, the picture frame and the bias extension tests can also be used for characterizing the forming behavior of fiber-matrix semi-finished products. In this case, they are heated to forming temperature. Differences in deformation behavior compared to dry fiber structures result mainly from the fact that fluid flows are induced during the movement of the rovings. The forming forces are therefore highly dependent on temperature.

For the description of the forming behavior, the **bending behavior** of the fiberstructures is also relevant, meaning the quantifiable flexural stiffness but also other qualitative factors, such as the occurrence of defects due to bending. While the latter rather requires an experience-based decision, the bending stiffness can be determined through experiments. For this particular test, methods from the clothing industry have been adapted. One of the most common methods, due to its simplicity, is the cantilever method (category 1). Adhering to DIN 53362, a strip of the fiber structure is positioned on a horizontal plate and then pushed over an edge (within 10 s) until the tip hangs down so low that an angle of 41° 30' is reached. The overhang length is then measured (Fig. 5.6). The stiffer the sample is in relation to its weight, the longer the overhang length. Thus, the bending stiffness can be calculated. Meanwhile, there are automated variants of this method, which are supposed to minimize the subjective influence, and other, more elaborate experimental setups have been developed, which will not be discussed here.

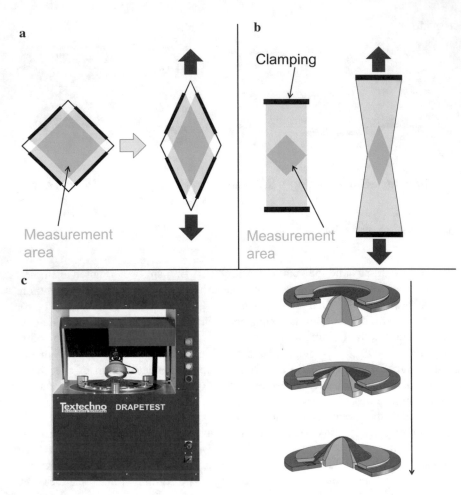

Fig. 5.5 Test procedures to investigate the draping behavior of a textile reinforcement; **a** picture frame test, **b** bias extension test and **c** "Drapetest" developed by the company Textechno. Images adapted, printed with permission of Textechno Herbert Stein GmbH & Co. KG

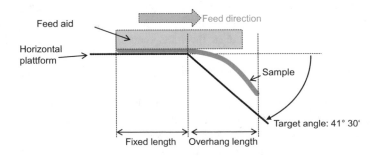

Fig. 5.6 Cantilever test used for the determination of bending stiffness of textiles

Compaction of semi-finished products with continuous fibers takes place in almost all processes where they are used. The main objectives are shaping, setting the target fiber volume content and, in the case of fiber-matrix semi-finished products, driving out air inclusions. On the one hand, knowing the forces required for compaction is essential for the selection of production systems. On the other hand, there are processes such as vacuum infusion, in which the achieved fiber volume content corresponds to the compacted state under atmospheric pressure and is therefore directly dependent on the compaction behavior. The compaction behavior itself results from numerous effects. The effects relevant for dry textiles are shown in Fig. 5.7 (right). The interaction of these effects typically leads to a correlation between compaction pressure and fiber volume content, as illustrated by Matsudaira and Qin [22]. This curve is a direct result of the dual-scale porosity. At the beginning of the compaction, the pore space between the filaments (microscale) and between the yarns (macroscale) is compacted. This is followed by microscopic fiber deformations, which lead to a linear increase. During decompaction, the model of Matsudaira and Qin shows a hysteresis. However, they considered fibers from textile production (polyester, cotton, etc.) [23]. In the case of glass or carbon fibers, which when subjected to lateral compression show a purely elastic behavior until failure [24], it is not to be expected that this phase will be reached in typical FRP production processes. Yet, a quasi-plastic deformation (within the time periods relevant for processing) can possibly be found in the second decompaction phase, because of macroscopic deformations of the fiber structure. This is, e.g., shown in [25], but also in various other studies. Nesting, i.e. the space-saving sliding of undulations of adjacent layers into each other, plays an important role here. The modeling for a process simulation therefore requires a viscoelastic consideration, as presented in [26].

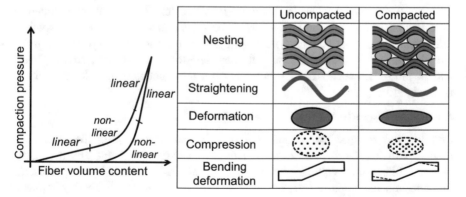

Fig. 5.7 Correlation of compaction pressure and fiber volume content (left). Adapted from Matsudaira and Qin [22], effects of textile compaction (right), adapted from Chen et al. [27]

Commercially available universal testing machines are used to determine the compaction behavior, whereby a sample or a sample stack of the fiber structure is compacted between two plane-parallel plates. Viscoelastic effects can be investigated by varying compaction speeds and holding times at different points, etc. By saturating the samples with a fluid, the influence of resin saturation can be simulated. Just as with shear, the presence of a matrix polymer has a very strong effect on the compaction behavior, as it induces fluid flows and affects frictional effects [26]. Even though the basic idea of the experiments is relatively simple, a first international benchmark study showed that the results of different research institutes can differ considerably [17]. The inherent deformation of the universal testing machines was identified as an important influencing variable, which has enormous effects given the small strains. The study also showed that the exact test methodology and test setup can differ significantly. It is therefore worth looking at the relevant technical literature before carrying out the corresponding tests.

Finally, with regard to the forming behavior, the **friction** behavior of the fiber structures is relevant. Of course, friction effects within the individual layers and the roving already play a dominant role concerning compaction, shearing, etc. Therefore, the frictional forces are implicitly considered in the forming effects described in the previous section. However, textile-to-textile friction is a relevant issue itself, as it inevitably occurs when a multilayer stack is formed. To increase process efficiency, stacks of flat textile layers are often first prepared and then jointly formed. With curved surfaces, however, this means that there are differences in radius over the cross section, which must be compensated by relative movement of the layers to each other. A lack of slippage resulting from a high friction coefficient can lead to waviness and wrinkling. For automated forming also friction between textiles and forming tools is relevant, as it determines the forming forces and has a significant influence on any defects that may occur. To determine the friction behavior, many investigation methods have been adapted from clothing industry. The main problem in this context is that textiles cannot be regarded as rigid bodies, but on the contrary exhibit strong deformation mechanisms, which in turn affect the contact surface. The description of the friction behavior, e.g. using the friction law according to Coulomb, can therefore lead to large errors. This is why no simple procedure can be shown here. A more detailed consideration of this is provided by [28].

Processing Viscosity of Molding Compounds

Compression molding compounds are pressed into form during the production process, causing the material to flow. This flow behavior is defined by the matrix polymer, fillers, fiber volume content and fiber lengths as well as by the processing pressure and temperature (preheating and mold temperature). With thermoplastic compounds, the tool contact initiates a cooling and therefore an increase in viscosity. Contrary, viscosity of thermoset compounds results from an interaction of rheological and reaction kinetic (cross-linking) effects and is thus complex in its time and temperature dependence. Here the tool contact initiates a heating, which at first reduces the viscosity. To characterize the viscosity behavior, press rheometers

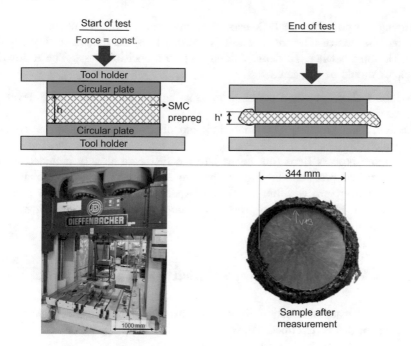

Fig. 5.8 Schematic procedure of the determination of the flow behavior of SMC with a press rheometer (top), press rheometer mounted in a press (bottom left) and example of a sample after the test (bottom right) [29]. images adapted, printed with permission of Leibniz-Institut für Verbundwerkstoffe GmbH

are used, which measure the behavior under processing conditions. Thereby, a compound sample is pressed under defined conditions and a one- or two-dimensional flow is generated. Figure 5.8 shows a schematic illustration of a press rheometer, used to determine the flow behavior of thermoset SMC molding compounds. Here, a SMC lay-up is positioned between two circular die plates in a press so that the plates are completely covered. Then the distance between the plates is reduced at a defined closing speed, and the press force is recorded.

In addition to the above-mentioned properties, depending on the application, others can be relevant for the selection procedure, but also for process modeling (e.g. thermal conductivity). However, these will not be considered in detail here.

5.2.1.5   Procedure for the Selection of Semi-finished Products

The preselection made considering Table 5.1 can now be further specified taking into account the issues raised in the previous section as well as the processing properties. Note that very often there is a trade-off between design, manufacturing and economic demands. A semi-finished product that offers excellent mechanical properties can be very challenging in processing and expensive in procurement. Therefore, semi-finished product selection in particular requires a procedure respecting the principles of IPD, especially involvement of all departments. The team member responsible for this task is the one from the manufacturing department. He must balance the demands and prepare a decision basis. The following principles should be respected:

1. Mature technologies and common material combinations reduce the development risk.
2. Design, manufacturing, material and economic demands must be equally considered.
3. The more process steps (on the way to the FRP) are already included in the semi-finished product, the higher the process robustness of the final FRP manufacturing process, but the more expensive the semi-finished product. Hereby, the FRP production process will not necessarily become correspondingly cheaper.

5.2.2 Selection of the Polymer Matrix

To finalize the material selection, the specific matrix polymer must be selected. The procedure for this is described below.

5.2.2.1 Procedure for the Polymer Selection

In the concept/draft phase, it was determined whether a thermoset or thermoplastic polymer is to be used. Now, the detailed selection of a matrix polymer is made. The market for polymers is extremely diverse and requires very specific expertise, which will not be transferred at this point. Instead, a simplified description of the selection process will be given. Subsequently, some important polymer types that are frequently used in the field of FRPs will be briefly introduced.

For the matrix polymer selection, the requirements defined in the requirements catalog as well as the requirements resulting from the process and semi-finished product selection are relevant. Figure 5.9 suggests a selection method for the matrix polymer. The method follows the basic principle that polymer systems already applied in the customer's company have the highest preference. Furthermore, the use of an industrially common standard system is preferable over a newly developed system [30]. The first approach in the selection process is therefore always to check if polymers (of the desired polymer class, i.e. thermoset or thermoplastic) are already in use and if so which ones. Selecting an already used system for a new product has several advantages:

- There is no need to search for a new supplier.
- Storing is simplified.
- There is experience in processing.
- Mechanical properties are already available for the design tasks, so possibly tests can be saved.

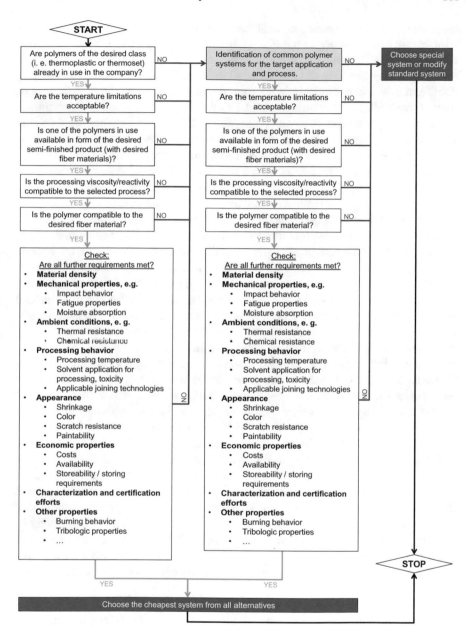

Fig. 5.9 Selection procedure for the polymer matrix

If one or more systems are available, the first step is to check whether the temperature limits comply with the requirements. This is often critical and therefore usually most suitable for making a preselection. The selection can be further narrowed down by checking the market availability of the desired semi-finished product form, which can also include fiber-matrix semi-finished products. If there is no market availability as a common standard semi-finished product, high additional costs for individual production can be expected. This would only be meaningful in the case of very large purchase quantities and corresponding series sizes. Of course, it can also be useful to check again at this point, whether another semi-finished product form can be selected, for which the desired polymer system is commonly available. However, if the market availability is given, all remaining alternatives are compared with the other requirements from the requirements catalog. Hereby, it is very important that all requirements from all departments are taken into account. Figure 5.9 shows the most important requirements that should be checked. Of all the alternatives that pass this evaluation, the cheapest one should be chosen. Generally, it should be taken into account that outstanding properties in one area are usually accompanied by deficiencies elsewhere [30]. Over-fulfillment of individual requirements that do not bring any relevant additional benefit can therefore make it more difficult to fulfill other requirements. It is therefore often advisable to first check the usability of resin systems with a balanced property profile.

If there is no in-house system available that meets all requirements, the next step is to try and find a common standard system for the desired application. Table 5.9 lists common matrix polymers. This table can be used to identify various suitable base polymers, for which the temperature limits can be checked first. Then the market availability of the specific semi-finished product form should be checked again. Here too, an alternative semi-finished product form may be useful if the market availability of the desired polymer is not common. Finally, a comparison with the requirements catalog is again carried out and, if this is positive, a balanced and cheap matrix system is selected. Literature, database or data sheet information can be used for the comparison. In the case of the latter, however, it should be noted that the mechanical properties are usually determined on climatically conditioned samples. It must therefore be ensured that the mechanical properties can actually be achieved under the planned operating conditions. Table 5.9 contains examples of some typical properties for frequently used polymers. For more detailed information, please refer to the relevant technical literature or the manufacturer's specifications.

If the requirements cannot be achieved with a standard system, the last alternative is to use a nonstandard system, which may have to be specially manufactured by the supplier. Alternatively, a standard system can be modified. At this point at the latest, it makes sense to contact potential suppliers. For many standard systems, additional modifiers can be purchased which allow the processing properties (e.g. reactivity and processing viscosity for thermosets), the mechanical properties (e.g. toughness) and other properties (e.g. flame resistance) to be influenced. However, this modification requires appropriate expertise, which is mainly present at the manufacturer, who only provides very limited detailed information on the exact

Table 5.9 Properties of standard selected polymers (Data partially from [30, 31–33])

Class	Polymer	Acronym	Density in g/cm^3	HDT-A^2 in °C	Tensile properties			Cost tendency
					Modulus in GPa	Strength in GPa	Elongation at break in %	
Amorphous thermoplastics	Polyarylethersulfone	PES	1.37	195	2.6–2.8	85	20–40	High
	Polyetherimide	PEI	1.27	197	2.9–4.5	95–105	60	High
Semi-crystalline thermoplastics	Polypropylene	PP	0.90–0.91	55–70	1.3–1.8	30–40	> 50	Low
	Polyamide 6	PA6[a]	1.12–1.15	65	1.5–3.0	80	> 50	Intermediate
	Polyamide 66	PA66[b]	1.13–1.20	75–100	3.0–3.5	75–100	> 20	Intermediate
	Polyetheretherketone	PEEK	1.32	140–155	3.5–3.8	90–105	> 50	High
Thermoset	Unsaturated polyester resin	UP	1.25–1.30	80–140	2.8–3.5	40–75	1.3–3.3	Low
	Vinyl ester resin	VE	1.1	100–150	2.9–3.1	80	3.5–5.5	Low
	Epoxy resin	EP	> 1.16	50–175	2.8–3.4	45–85	1.3–5.0	Low to intermediate

[a]Dry
[b]Heat deflection temperature

chemistry. If no data are available for a selected system, for example, because it has been specially adapted to the application by modifiers, these must be determined. Corresponding procedures are presented in the following section.

5.2.2.2 Thermal Analysis of Polymers

The mechanical properties of polymers strongly depend on the temperature. Hence, the methods of thermal analysis are often applied. Here, testing is carried out on conditioned test specimens in order to detect property changes due to environmental influences and ambient conditions. For polymers, e.g. the moisture content is relevant, which is why conditioned samples are tested at specific temperatures and moisture contents. Depending on the application, further specific conditioning is common (e.g. UV radiation or salt water). The most important methods of thermal analysis are explained below. Further information can be found in [34, 35].

Dynamic thermo-mechanical analysis (DTMA) refers to the measurement of the mechanical response under slight oscillating loads. Changes of the mechanical properties of interest are measured with regard to time, temperature and/or frequency. This way, the design-critical deformation behavior (Fig. 4.55) as well as, e.g., the glass transition point can be determined. Possible loadings are tension, bending, compression and shear.

In the **differential scanning calorimetry (DSC)** the temperature and heat flow associated with a phase change in the material are determined as a function of time and/or temperature, under controlled ambient conditions. DSC thus serves to determine the glass transition temperature, the melting point, the decomposition temperature, the degree of curing, the specific heat capacity, the crystallinity, the melting and crystallization behavior, the reaction kinetics, the oxidative/thermal stability and the effectiveness of additives.

In **thermogravimetric analysis (TGA)**, the change in mass is measured as a function of temperature or time. This provides information on outgassing, decomposition, oxidative processes, inorganic filler content and in principle also fiber volume content (via ashing).

In addition to the properties of the final component, the **processing properties** of matrix polymers are particularly relevant, especially with regard to the impregnation process. Therefore, the temperature and time-dependent fluid viscosity must also be characterized. The viscosity behavior of thermoset resin systems is determined by the interaction of rheological and reaction kinetic (cross-linking) effects. It is very complex in terms of time and temperature dependence [36] and usually determined by means of commercially available shear rheometers. Figure 5.10 shows the typical viscosity development of a thermoset resin system that is injected into a mold with a higher temperature than the resin. Initially, the temperature increase leads to a reduction in viscosity (rheological effect), but the increasing cross-linking results in an opposing, viscosity-increasing effect (reaction kinetic effect). This results in a "bathtub curve." The higher the mold temperature,

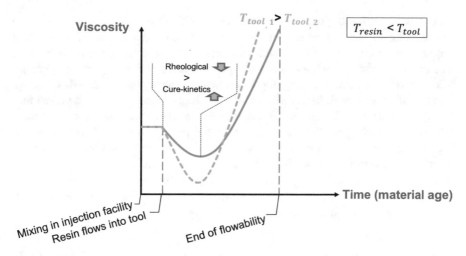

Fig. 5.10 Illustration of the viscosity development of a thermoset resin system being injected into a tool with elevated temperature

the greater the viscosity reduction at the start of injection, but also the faster the reaction and the faster the maximum flow viscosity is reached. From this point on, no remarkable additional flow front progress is achieved. If the component is not completely filled at this point, it is usually scrap. Finding the ideal combination of resin chemistry, preheating and mold temperature are therefore no trivial tasks.

In thermoplastics, shear rate-dependent effects are given in the form of a shear thinning of the melt due to an increasing orientation of the molecules in the direction of flow, which reduces the internal friction [37]. The higher the shear rate, the greater the viscosity reduction (eventually the reduction will stagnate). The viscosity is thus influenced by, among other things, the injection pressure, the resulting flow velocity, the component geometry and the structure of the fiber reinforcement. An exact quantification of the viscosity given in the process is therefore difficult, but the viscosity comparison of different polymer systems under the same boundary conditions can be used for selection.

5.2.3   Determination of FRP Material Properties

During concept and design development, literature or data sheet values of the material properties can be used. For detailed elaboration, however, exact, reliable mechanical values for the specific selected semi-finished products in combination with the desired manufacturing process are required. In particular, if a nonstandard material combination has been selected and therefore no mechanical properties are available, the selected materials/semi-finished products must be procured and test specimens must be produced. The process for manufacturing the test specimens should be as close as possible to the final manufacturing process, as this has a strong influence on the component properties. Even if all this is adhered to, it should always be taken into account that there are still manufacturing influences that are not covered by common test methods and specimen (e.g. the draping behavior of fabrics when adapting to curved or even double-curved surfaces). Further influences result from the ambient conditions. These too must be taken into account during the design of the experiment, as they can have an enormous influence depending on the material (e.g. if the matrix absorbs water). Hence, it becomes clear that there are several issues when determining the material characteristics. In the following section, the most important mechanical properties and corresponding test methods are briefly presented including some insights on particularities correlated to FRPs. This transfers the basic knowledge required for each team member during IPD with FRPs.

Tensile Strength/Stiffness

The tensile properties of a FRP are usually determined in a tensile test according to DIN EN ISO 527-5 (unidirectional-reinforced) or DIN EN ISO 527-4 (multi-axial-reinforced). In the tensile test, the stress-strain behavior is measured under quasi-uniaxial tensile load (continuously increasing load) until breakage. If fibers are parallel to the tensile direction, the tensile properties are strongly fiber-dominated, so that the fiber orientation plays a very important role. Therefore, the test can be used to make a statement about the quality of the fiber orientation by cutting test specimens from a test plate at different angles. However, due to the fiber dominance, statements on the quality of the fiber-matrix adhesion are only possible to a very limited extent, when fibers are present in the direction of tension. Anyway, a qualitative evaluation can possibly be made by an analysis of the fracture surface. According to a rule of thumb, poor fiber-matrix adhesion tends to result in extensive "fraying" of the fracture surface, i.e. there are many loose and relatively long fiber ends which have been pulled out of the matrix, while leaving corresponding holes in the matrix on the opposite side. When viewed with a scanning electron microscope (SEM), the fiber surfaces often appear smooth and there is little to no matrix adhesion. With good fiber-matrix adhesion, the fracture surface is usually comparatively "smooth," with no or only short extended fiber ends visible. In SEM images, strong matrix adhesion could be seen on the fiber ends. A special case is given by the test of unidirectional-reinforced test specimens, with the fiber orientation transverse to the tensile direction (transverse tensile test): Here, the matrix properties in combination with the fiber-matrix adhesion are decisive.

Finally, it should be noted that due to the comparatively small elongation at break of typical fiber materials, like glass and carbon, FRPs typically do not show a pronounced yield strength, but rather brittle failure.

Bending Strength/Stiffness

To determine the bending properties, three-point or four-point bending according to DIN EN ISO 178 or DIN EN ISO 14125 is applied. In the three-point bending test, a freely supported beam is loaded in the middle of the span width. There is a maximum load in the middle. In the four-point bending test, there is a constant bending moment between the supports.

Shear Strength

The shear strength of FRPs is determined according to DIN 53399-2, with a picture frame in a tensile testing machine or, in the case of shear laminates (±45° laminates), directly from a tensile test with rectangular test specimens according to DIN EN ISO 14129.

Interlaminar Shear Strength

Interlaminar shear strength (ILSS) is of particular importance for FRPs, as the risk of delamination is one of the greatest weaknesses, particularly for continuous-reinforced FRPs. Here, the short beam test according to DIN EN ISO 14130 is often used, where a freely supported beam is loaded in the middle of the span width. In comparison with the classic three-point bending test, the span is remarkably smaller (10 mm instead of 64 mm). The maximum shear stress is determined in the neutral fiber axis at the moment of failure (apparent shear strength). The test is very well suited to gain insights into the fiber-matrix adhesion, due to the dominant shear forces. However, the determined mechanical properties are rather suitable for quality control than for the generation of values for design. This is why they are also referred to as "apparent ILSS" [38]. DIN 65148 defines a method to determine the interlaminar shear strength in a tensile test. Here, above and below the intended shear zone grooves are milled into the sample in order to induce a shear load.

Compressive Strength/Stiffness

To determine the compressive strength, e.g. DIN EN ISO 14126 is applied, where a compressive load is generated either by shear stress induced by cap strips or via direct compressive stress on the end faces (bending strain in the test specimen must remain below 10%).

Impact Strength

Due to the low elongation at break of typical fiber materials such as glass and carbon, FRPs are relatively brittle. Hence, consideration of the impact strength is even more important. This is typically determined by the impact bending test according to DIN EN ISO 179 (Charpy impact strength), were a sample is positioned on two abutments and abruptly loaded with a single blow of a hammer.

Compression After Impact

If an impact occurs on a FRP component, this can lead to interlaminar damage, especially in the form of delamination. This can lead to a drastic deterioration of the mechanical properties, without this being visible from the outside. In order to be able to make a statement about the performance loss due to an impact, the residual compressive strength is tested according to DIN 65561. In this test, often referred to as compression after impact (CAI), a test plate undergoes a drop impact, which induces a defined pre-damage. Then the quasi-static residual compressive strength is determined in a testing machine.

Laminate Structure and Quality

To support interpretation of the results of the tests mentioned above, it makes sense to get a more precise image of the structure of the laminate, i.e. exact fiber volume content, fiber distribution, fiber orientation, textile deformation, etc. In addition, the laminate quality, for example, in terms of number, size and shape of possibly existing pores, can provide further insights. The first approach, due to comparatively little effort, is usually the creation of microsections and observation in a light microscope. Porosity, fiber volume content and textile deformations are already clearly visible here. Also, a first statement regarding the fiber orientation can be made; for example, a straight cut perpendicular to the direction of the fibers should form a circular cross section, while an elliptical cut surface indicates an orientation that is not perfectly perpendicular to the image plane. However, an exact quantification is not possible without the third dimension. The determination of the fiber volume content can only be done by relating the sum of fiber section areas to the total section area. Insufficient clarity at the boundary areas between fiber and matrix makes it difficult to assign them precisely. These inaccuracies can be increased by inexperienced sample preparation. Additionally, there is the inherent uncertainty of any statistical method that is intended to make statements about a whole sample volume based on very limited data (in this case a cut surface of limited size). This problem exists for all the information made based on microsection analysis.

A somewhat more detailed evaluation is possible by a computer tomographic examination (μCT), which allows a three-dimensional view, but is more complex and requires more in-depth knowledge for a meaningful analysis. The statistical problem remains, of course, especially since the effort for sample analysis (and the costs for the imaging system) increases with the sample size. In addition, a larger scan volume often requires a reduction in resolution. The safest way to determine the fiber volume content is therefore ashing or chemical dissolution of the matrix in combination with weighing the specimen before and after removal of the matrix.

Depending on the application, other, more specific parameters may become relevant, e.g. with regard to energy absorption in the event of damage. Overall, the above-mentioned mechanical properties already provide a good overall picture for a FRP and can thus form the basis for the elaboration of the design.

5.3   Elaboration of the Design Concept

During the elaboration of the design, the final properties of the product are determined. This means that the decisions made here affect all phases of the product life cycle to great extent. The design thus forms the core of the IPD and requires many more than any other task in the development, holistic thinking and the motivation to work in a team. It is the task of the team member from the design department to implement meaningful integration measures. This should especially include design reviews and clearance loops, as well as team meetings, where every critical decision is discussed. However, a truly successful IPD does not result from milestones, checklists and predefined meetings. The humans are the decisive factor! Those involved must share the objectives of the IPD. An informal exchange between a design engineer and a production engineer, working together for the benefit of the product, is much more effective than weekly meetings dictated "from above," which are perceived as an annoying duty. In short, the designer must be convinced from the benefit of the input of his colleagues. The following sections intend to contribute to this mind-set.

Designing components requires extensive expertise. This applies to all materials and especially to FRPs, due to their particularities. The objective of this book is not to transfer the detailed knowledge required for the design with FRPs to the team member entrusted with the design. For this, one can refer to corresponding books dealing with this specific topic (e.g. [30, 39–42]). Rather, this book intends to facilitate efficient communication within the IPD team, by mediating the basic knowledge for design with FRPs. Therefore, the following sections will show how during design the product life cycle as a whole can be taken into account. To do so, the aspects of a design to

- manufacture,
- join,
- repair, and
- recycle

are discussed.

5.3.1   Design to Manufacture

Design to manufacture includes two aspects: first, the consideration of the process-specific requirements so that a component can be manufactured efficiently or at all (e.g. draft angles), and second, the consideration of manufacturing-related deviations from the idealized design of a FRP, e.g. deviations from the idealized fiber orientation. Both aspects are often directly related to each other.

The consideration of process-specific requirements is often referred to as "design for manufacturability." It requires a precise knowledge of the processes, which is why cooperation with the team member from the manufacturing department is indispensable. If the process was carefully selected (see Sect. 4.6.3), the general manufacturability should already be given. At this point, the aim is to ensure that the design requirements and restrictions, resulting from the specific process, are considered. The following tables (based on the VDI Guideline 2014 [43]) give an overview of relevant requirements and limitations.

Table 5.10 shows the large differences between FRP manufacturing processes with regard to the (usual) maximum limits of wall thicknesses and their variation. For design engineers, it is also important to ensure that the process-related manufacturing tolerances, as listed in the table, are considered.

Table 5.10 Design requirements and limitations of different manufacturing processes for FRPs, adapted from VDI Guideline 2014 [43]—part 1

Processes	Design elements		
	Part thickness in mm	Thickness changes	Thickness variation (process-induced)
Fiber spraying	Typically 2–10	Possible (1:2–1:3)	Up to 80%
Winding	Vessels: > 4 Pipes: > 1	Possible	Up to 20%
Pultrusion	3–20	Possible for cross section	Up to 10%
Tape laying	Typically 2–6	Possible	5%
Autoclave	2–20	Possible	10%
Prepreg compression molding	Typically 1.5–6	Possible	1–3%
Thermoforming	Typically 2–8	Possible (low, 10%)	5%
Hand lay-up	Typically 1–10	Possible (1:2–1:4)	Mats up to 50% Woven fabrics up to 20%
Liquid composite molding	3–15	Possible	Closed tool up to 5%, open tool up to 20%
Compression molding	2–30	Possible	0.1% (for thermoset)
Injection molding	1–10	Possible	0.1%

Table 5.11 shows the requirements for draft angles. If these are not available or not sufficient, demolding can damage the part, e.g. by delamination. The component thickness also plays a role here, as it has an enormous influence on the bending stiffness. The table also lists minimum radii, which should be followed to ensure that they can be cleanly molded when using typical semi-finished products. Undercuts are possible for some processes, as shown in the table. However, they always lead to increased effort, especially with regard to tool design and loading. This is because a parting plane is usually required to enable loading and demolding.

Table 5.12 lists geometric particularities that cannot be achieved with all processes. This includes ribs, which are difficult to process with textiles but quite easily manufacturable with molding compounds. Furthermore, it is shown which methods allow the integration of metallic inserts or opening.

Deviations from the boundary conditions mentioned in the tables can have a strong impact on the technical manufacturability and the cost-efficiency. However, the boundary conditions also serve to avoid process-related production errors or deviations from the design specifications. Nevertheless, such errors cannot be completely prevented. Therefore, they have to be taken into account in the design

Table 5.11 Design requirements and limitations of different manufacturing processes for FRP, adapted from VDI Guideline 2014 [43]—part 2

Processes	Design elements		
	Demolding angle	Minimum radii (in mm)	Undercuts
Fiber spraying	1°–2.5°	5	Possible
Winding	–	10	Conditionally possible
Pultrusion	–	1	Possible perpendicular to pulling direction
Tape laying	1°–2.5°	> 20 (determined by diameter of roll)	Conditionally possible
Autoclave	1°–4°	3	Possible
Prepreg compression molding	0.5°–1°	Inside 0.8/outside 0.2	Possible
Thermoforming	0.5°–1°	3	Difficult
Hand lay-up	1°–2.5°	5	Possible
Liquid composite molding	1°–4°	5	Possible
Compression molding	0.5°–1°	> 0.5	Possible
Injection molding	0.5°–1°	> 0.5	Possible

by applying realistic safety factors. In the following section, a short introduction to this topic is given.

Probably the most important topic in this context is deviations of the intended and the actual fiber position and/or orientation. These can have various causes. A large number of variations can already result from the semi-finished fiber product. Apart from the fact that, for example, the mechanical properties that can be achieved with a fabric structure are limited by undulation, manufacturers typically state variations concerning the areal weight of ± 5%. It is therefore essential to determine the mechanical properties on samples made of the final semi-finished products. As long as these mechanical properties are not available, a conservative design is necessary in order to prevent costly redesigns at a later stage. More influences arise during the further processing of the fiber structures. This includes, e.g., molding errors when depositing the fiber layers on contoured tools. As illustrated in Fig. 5.11, such defects can occur at edges, for example, especially when thick individual layers or multilayer stacks are deposited. At the same time, edges are often critical structural areas in which the achievement of the required mechanical properties is of utmost relevance.

Table 5.12 Design requirements and limitations of different manufacturing processes for FRPs, adapted from VDI Guideline 2014 [43]—part 3

Processes	Design elements		
	Ribs as stiffeners	Metallic inserts	Openings
Fiber spraying	Conditionally possible	Possible	Possible
Winding	Conditionally possible	Possible, typically retrospectively	Not possible
Pultrusion	Possible perpendicular to pulling direction	Not possible	Not possible
Tape laying	Not possible	Possible, typically retrospectively	Possible
Autoclave	Possible	Possible	Possible
Prepreg compression molding	Conditionally possible	Possible	Possible
Thermoforming	Conditionally possible	Possible	Retrospectively possible
Hand lay-up	Possible	Possible	Possible
Liquid composite molding	Conditionally possible	Possible	Possible
Compression molding	Possible	Possible	Possible
Injection molding	Possible	Possible	Possible

Such placement defects must be prevented by means of process design, e.g. by a suitable selection of semi-finished products (thin layers facilitate sliding) and corresponding lay-up strategies (single layers with intermediate compaction instead of multilayer stacks). It is even more important to reduce the risk of such defects in the first place, by means of geometric design, e.g. by respecting minimum radii adapted to the process and the semi-finished products. For this, rules of thumb exist, which, for example, define minimum radii: These are between 0.2 and 0.5 mm for GFRPs and between 0.5 and 1 mm for CFRPs [44].

Another important point is given by intentional and unintentional draping effects (see Sect. 5.2.1.4). Figure 5.12 shows an example of a woven and a non-crimp fabric, which first sheared to the maximum shearing angle and then back to original orientation again. The pictures show that this process causes irreversible deformations of the textile structure.

When processing fiber structures, they have to be pulled from rolls, stacked and cut. Draping in the form of shearing and stretching cannot be fully prevented when doing so, even if the fiber structure is of high reproducible quality and the

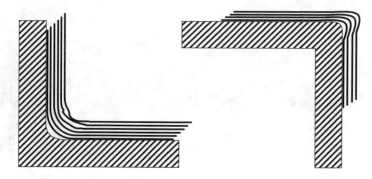

Fig. 5.11 Placement defects when draping multiple layers at edges

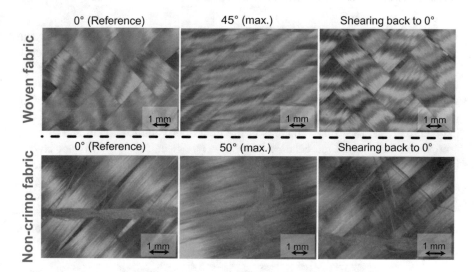

Fig. 5.12 Lasting effect of shearing on a woven and a non-crimp fabric. Printed with permission of Leibniz-Institut für Verbundwerkstoffe GmbH

preparation is minimally invasive. At the latest during further processing into a composite, the effects of fiber reorientation are difficult to predict. This applies to the forming of both dry and pre-impregnated semi-finished products. When textiles are adapted to a three-dimensional structure, shearing inevitably occurs. Simulation programs can provide a prediction of the fiber orientation in the formed component. However, these are usually based on purely kinematic models (Fig. 5.13).

A slippage of fibers, as shown in Fig. 5.14, can usually not be predicted by corresponding simulation programs. If such defects occur during serial production, a high reject rate can be the result, making the component economically inefficient. This can be counteracted by avoiding too steep deformations.

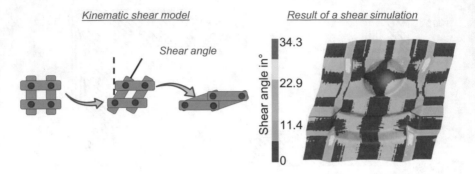

Fig. 5.13 Kinematic model and kinematic-based simulation for the prediction of shear angle distribution resulting from the draping of a woven fabric. Right image [45], printed with permission of Leibniz-Institut für Verbundwerkstoffe GmbH

Even if there is no fiber deformation of the single layers, other effects can nevertheless exist, which affect the mechanical performance at the component level. A typical example is a locally varying fiber volume content, as it can occur, e.g., in the vacuum infusion process, in which the fiber structure is compacted by atmospheric pressure. As shown schematically in Fig. 5.15, differences in the nesting behavior, caused by different orientations of the individual layers in relation to one another, can lead to strong variations in the final fiber volume content and thus the component weight and the bending properties (see Sect. 5.2.1.4). It is not always possible (or cost-efficient) to avoid these effects by means of design or process optimization. It is therefore important that the designer is aware of such variables, when designing the component.

Fluid flows can also cause fiber disorientation (so-called fiber washout), especially at high injection pressures. Figure 5.16 shows the fiber deformations on a

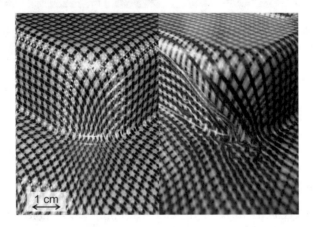

Fig. 5.14 Fiber sliding on a formed organo sheet. Printed with permission of Leibniz-Institut für Verbundwerkstoffe GmbH

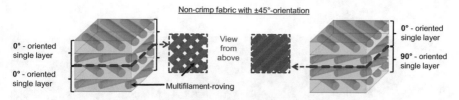

Fig. 5.15 Different nesting behaviors of two identical, but differently stacked biaxial non-crimp fabrics [46]. Printed with permission of Leibniz-Institut für Verbundwerkstoffe GmbH

unidirectional fabric resulting from a ring injection with 3 bar injection pressure. The textile was pushed together in such a way that the increase in fiber volume content blocked the fluid flow inward, so that ultimately no complete impregnation was achieved. In the case of short fiber compression or injection molding compounds, the flow can induce a fiber alignment, which can lead to an anisotropy in the material behavior, which is not taken into account during design [47]. Figure 5.16 shows an example of flow-induced fiber alignment in an injection molding process.

Further errors can occur when the fiber structure and the matrix are brought together in the process or when pre-impregnated semi-finished products are laminated and consolidated.

Very often, the **pore content**, the volume fraction of pores in the laminate, is used as a criterion for the quality. In the aerospace industry, for example, there are usually strict limitations for the pore content, as pores can lead to local stress peaks due to their notch effect and thus form the starting point for cracks. This is particularly relevant concerning fatigue [30]. Pores result from air inclusions or outgassing from the polymer. Figure 5.17 shows an example of a microsection of a CFRP plate produced by prepreg compression molding.

Fig. 5.16 Wrinkling of a preform resulting from fluid flow during ring injection (left, printed with permission of Leibniz-Institut für Verbundwerkstoffe GmbH), cross section of an injection molded glass fiber-reinforced polymer—flow-induced fiber orientation is clearly visible (right, printed with permission of InnCoa GmbH)

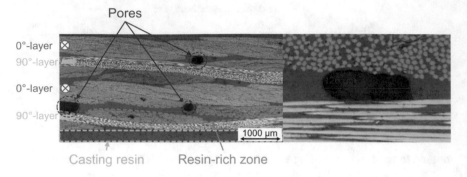

Fig. 5.17 Microscopic cross section of a CFRP (woven fabric in epoxy resin) at about 54% fiber volume content—pores within the laminate are clearly visible (right image shows a close-up)

The risk of pore formation very much depends on the process technology. Therefore, if a high laminate quality is required, the process should be selected accordingly (see Sect. 4.6.3). It is important to respect the limitations resulting from the selected process from the beginning on, in order for the processing to be cost-efficient.

In addition to the actual production of the FRP component, also the post-treatment can have a strong impact on the component properties. It is known that the mechanical processing of FRPs by drilling, sawing, milling, etc., is challenging due to the heterogeneous material structure. It can induce thermal damage to the matrix, as the thermal conductivity is comparably low and coolant can lead to moisture absorption by the matrix [48]. Mechanical processing also goes together with the risk of delamination, as shown in Fig. 5.18 for drilling and sawing. During drilling, the twist of the cutting edge generates forces, which can peel the top layer from the laminate. Another critical moment is when the drill pushes on the bottom layer [49]. Defects resulting from drilling are often particularly critical, since forces are often applied to drill holes (e.g. via screw connections).

Alternative cutting methods, such as laser cutting or water jet cutting, also provide challenges, especially the risks of thermal damage, delamination and moisture absorption. Since the mechanical processing of FRPs is generally considered challenging, it is all the more important to think about required cutting lines and holes in the part from the beginning on. Perhaps, design measures reducing the risks can be identified, e.g. moving cutting lines in less critical areas.

In addition to the examples mentioned, there are of course other effects that can be highly relevant, such as spring-in effects in part forming [50–52]. Overall, the following conclusions should be kept in mind during design elaboration:

1. Each manufacturing process provides specific **design restrictions**. Ignoring these restrictions, in the worst case, causes the part to be unmanufacturable. Surely, economic efficiency will be reduced, while the risk of process-related manufacturing errors is increased. This must then be taken into account during

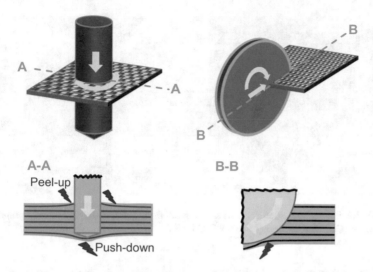

Fig. 5.18 Danger of delamination during drilling (left) and sawing (right) of continuous fiber-reinforced polymer

design by means of increased safety factors, which again reduces functionality (especially lightweight potential). In cooperation with the colleague from the manufacturing department, the limitations must therefore be discussed before design elaboration, but also in the context of regular design reviews.

2. **Process-related deviations** from the ideal structure are never fully avoidable, even if a perfectly suitable process was selected and all design limitations were respected. Therefore, the component design must be considered specifically with regard to expected deviations, in order to take into account uncritical deviations through adapted material characteristics and to counteract critical deviations through design adjustments.

5.3.2   Design to Join

When designing a FRP component, it must be taken into account that perhaps it must be integrated into a superordinate structure. It must therefore be designed according to the requirements resulting from a possible joining process. In general, as shown in Fig. 5.19, a wide variety of joining methods can be used.

When using these methods however, a whole series of FRP-specific particularities must be taken into account. The most important ones will be briefly explained in the following.

Thermal expansion mismatch: FRPs, especially CFRPs, typically show significantly lower coefficients of thermal expansion (CTE) compared to metals.

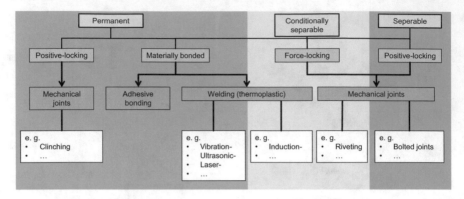

Fig. 5.19 Joining technologies for FRPs. Adapted from [107] (Image adapted, printed with permission of Carl Hanser Verlag GmbH & Co. KG)

Therefore, if they are joined with metals, stresses occur during joining when heated as well as during subsequent cooling. Even later when the part is in use, varying ambient temperatures can lead to different thermal expansion or corresponding stresses, which can ultimately lead to failure of the joint. This topic is becoming increasingly important, as multi-material design following the paradigm "the right material in the right place" is gaining importance. Problems caused by CTE mismatch can be counteracted, for example, by joints that are flexible to a certain extent, such as appropriately thick adhesive bonding films, or, alternatively, by selecting a suitable FRP. For example, classic glass fiber SMC has an almost identical CTE to steel [54].

Contact corrosion: If CFRP is joined with aluminum or steel, there is a high risk of contact corrosion. The reason becomes clear, when the potential differences between the graphite present in the carbon fibers and the metals are considered (Fig. 5.20). Theoretically, the carbon fiber in the FRP is completely surrounded by the matrix, which would prevent corrosion of the metal. However, the production of the joint can lead to exposure of fibers and thus to direct contact.

Accordingly, design measures must be taken to prevent contact between fibers and metal. This can be achieved, for example, by subsequent sealing of mechanically processed CFRP edges or by the integration of glass fiber cover layers. The selection of the joining method is also important. For example, an adhesive bond offers the possibility to create a separating protective layer [57].

Creep behavior: Thermoplastics tend to creep during long-term loading, which leads to a loss of tension in pretensioned rivet or screw connections [58].

Surface energy: Thermosets can be adhesively bonded by epoxy resin or polyurethane very well. However, nonpolar thermoplastics are comparatively difficult to bond, which is related to the low surface energy. This applies to many semi-crystalline thermoplastics. Therefore, they must be prepared for bonding, e.g. by plasma activation [59, 60].

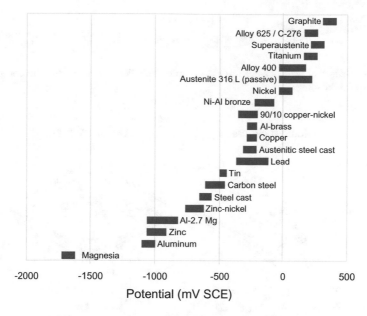

Fig. 5.20 Potential differences between graphite and different metals in seawater [55] after [56]. Image adapted, printed with permission of IBEKOR GmbH

Shear adhesive bonding: The most critical load case for an adhesive bond is one that leads to peeling. The load should ideally be designed for pure shear. Stress peaks should be avoided, for example, by stepped joining. It must also be taken into account that the load causes a deformation of the joint, which is why a symmetrical structure may be appropriate (Fig. 5.21).

FRP-suitable load introduction: Screwed connections are usually not considered suitable for FRPs, due to the required drilling, which locally impairs the structural integrity. Still, screwed connections can be found quite often in the field of FRPs, simply because they usually represent an economically advantageous solution. It is important to counteract the local weakening, e.g. by local adaption of the laminate with additional layers, as shown in the example given in Fig. 5.22.

Figure 5.23 shows a load introduction element that can be sewn on top of a textile preform, or by using additional cover layers, also into a lay-up. By means of a resin injection process the element becomes an integral part of the FRP. As the figure shows, during a pullout test, a strength can be achieved that leads to a primary failure in the metal component and not in the FRP. Obviously, such load introduction elements are quite complex in terms of processing. In general, it can be said that a higher technical value of a solution usually goes hand in hand with increasing costs [61, 62].

Heat input via the joining partner: The temperature resistance of the polymer matrix will usually be below that of a potential metallic joining partner. If the metal part borders on areas where high temperatures are expected, it must be noted that

Fig. 5.21 Danger of peak stresses resulting from deformation at single (left) and double (right) shear adhesive bonding

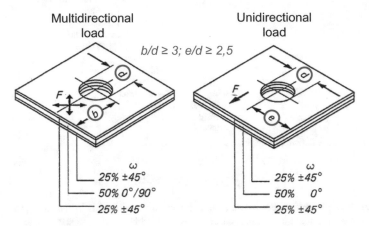

Fig. 5.22 Adaption of the laminate to the load situation [43] (Image adapted, reproduced with the permission of the Verein Deutscher Ingenieure e. V.)

Fig. 5.23 Load introduction element sewn into textile preform (left), after resin injection (middle) and after pullout tests with failure visible on the metallic component (right). Images adapted, printed with permission of Leibniz-Institut für Verbundwerkstoffe GmbH

the high heat conduction of the metal may lead to heat introduction into the joining zone and into the FRP. Accordingly, the joint and the FRP must also be adjusted to the corresponding temperature ranges, or a sufficient thermal insulation must be incorporated into the joint.

Overall, it should become clear that the design of joints requires special knowledge and the assembly conditions must be clarified at an early stage.

5.3.3   Design to Repair

IPD should consider the subsequent steps not only of product creation, but also of the life cycle phases beyond. This of course also includes the use phase. During use, the component can be damaged. This must be addressed with a repair strategy. In order to achieve a design suitable for repair, it must first be evaluated which forms of damage can occur during use and how critical these forms are with regard to the component functionality. This can be done, for example, by means of a failure mode and effects analysis (FMEA) (see Sect. 5.4.3). The spectrum of defects can range from scratches and fiber breakouts to delamination and punctures. Chemical or thermal damage, e.g. from lightning strikes, is also possible. Functionality can refer to load-bearing functions, but also, e.g., to optical functions. Taking into account the probability of occurrence and the damage potential, it must then be decided, how the damage risks are addressed in terms of design. There are basically four possibilities to deal with possible damages, and one of them must be selected for each sufficiently likely damage type:

1. **Prevent** damages by means of design.
2. Ensure **tolerance** to the damage by means of design.
3. **Repair** the damage.
4. **Replace** the damaged component.

These possibilities will be examined in more detail below.

If a possible damage is so critical that a repair would not be feasible, neither technically nor economically, the design must provide for the possibility of **replacement**. This must then be taken into account, when selecting the joining technology.

Providing the possibility for **repair** is challenging with FRPs, due to their specific structure. Superficial scratches up to exposed fibers can be re-sealed. However, if delamination occurs, it becomes difficult to restore the original cohesion. If necessary (for thermoset components) resin can be injected into the damaged area, provided a limited size. If the damage extends deep into the structure, the problem must be solved that separated fibers cannot be rejoined. The damaged area must therefore be bridged. One way to do this is to remove the damaged area and bridge it on both sides with a riveted repair piece. However, this is accompanied by an increase in weight. Furthermore, the surface is no longer flat. In terms of mechanical performance, it is better to first stepwise remove the FRP around the damaged area (Fig. 5.24). Then cutout FRP layers (prepreg or cured) can be placed in the area and bonded or cured. This way, the strength can be restored and the surface is flat again. [63]

The repair methods are comparatively costly and therefore often not economic, which is one of the reasons that ultimately lead to a decision against the use of FRPs.

If neither repairing nor replacing the damaged part seems meaningful, the part must be designed in a way that the possible occurring damages do not impair the

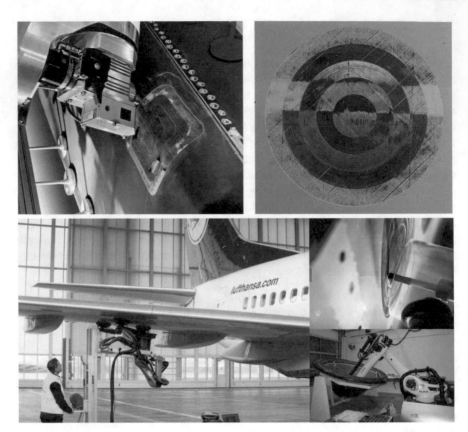

Fig. 5.24 Scarfing of airplane component (top images, [64] printed with permission of Springer Fachmedien Wiesbaden); self-carrying milling robot for stepwise and continuous scarfing. Bottom images, printed with permission of Lufthansa Technik AG

functionality. That is, either the component is designed to be **damage-tolerant** or the **probability** of damage occurring is **reduced** (e.g. by shielding). Both options usually affect the lightweight potential. Which of the strategies makes sense from a techno-economic point of view depends on the respective industry. A corresponding overview of repair strategies for FRP structures in aircrafts is, e.g., given in [65].

The design for repair also includes the aspect of **damage detection**. Depending on the possible damage, an appropriate repair strategy must also define the methods to be used for damage analysis (see Sect. 5.4.3). This might result in additional design requirements, such as a possibility for two-sided access.

5.3.4   Design to Recycle

In its directive 2008/98/EC on waste ([66], 19.11.2018) the European Parliament and the European Council defined the objective of a European recycling society, with a high level of efficiency in the use of resources. In the long term, therefore, a broad usage of FRPs can only succeed if appropriate strategies for resource efficiency are developed for this group of materials. This also requires to think the basic idea of IPD to the end and to consider all phases of the product life cycle, including end of life and aftercare, from the beginning of product development on. This is because the possibilities for recycling are widely determined during the design phase—either consciously or unconsciously.

Different types of waste emerge in the field of FRPs. First, components that have reached the end of the product life cycle due to aging or irreparable damage are referred to as end-of-life (EoL) components, and second, so-called in-house waste that is already generated during production. This includes dry waste, such as roving spool remnants and textile scrap, which arise during the processing of dry semi-finished products. Furthermore, in-house waste includes impregnated waste, such as thermoset prepregs, which are no longer usable due to aging, thermoset or thermoplastic prepreg scrap or waste from mechanical post-processing (milling dust, edge trimmings), and finally rejected parts. By 2020, about 20,000 t of CFRP components are to be recycled, which corresponds to about 10% of the current annual production volume. These components contain about 12,500 t carbon fiber. The total amount of fibers to be recycled can be approximately divided into 1/3 in-house waste (offcuts, rejects, etc.) and 2/3 EoL components [67–69].

Besides the general motivation to develop a recycling strategy for economic reasons, to improve the company image or to respect legislation, there are also FRP-specific reasons to pay attention to this topic. For example, the production of carbon fibers is very energy-intensive and associated with corresponding costs. At the same time, the waste rates are very high, especially for textile reinforcements. Finally, "resource efficiency through lightweight design" is an important sales argument for FRP components, which clearly loses impact when lacking a recycling strategy. Hence, the question arises of how to deal with the waste. The aforementioned EU Directive [66] establishes a hierarchy (order of preference) for the handling of waste. It is shown in Fig. 5.25.

For each of the shown stages examples are given below with the intention to provide the reader with food for thought concerning design to recycle with FRPs.

Avoidance: The best strategy to deal with waste is to avoid it in the first place. In the context of FRP production, this can be realized as follows:

- Use of material-efficient manufacturing processes. For example, the use of semi-finished products from a tape-laying process instead of textile-based semi-finished products to reduce scrap.
- Correspondingly optimized component design, which minimizes edge trimming.

Fig. 5.25 Waste hierarchy corresponding to recycling management act [70]

- Use of a robust manufacturing process that minimizes the rejection rate where appropriate, eventually by accepting deficiencies elsewhere (e.g. in terms of cycle time).
- Use of locally defective components as "donors" for the repair of components in use.
- Sophisticated storing concepts combined with a corresponding production planning. This is particularly important for semi-finished products with limited shelf life (especially thermoset prepregs), in order to minimize waste rates.

Reuse: EoL waste is avoided if the EoL components are simply reused for the same purpose without extensive pre-treatment. This generally corresponds to the principle of reusable bottles for beverages, which are used repeatedly for exactly the same purpose (until eventually they are critically damaged). An application of this concept to FRPs is generally conceivable. Figure 5.26 shows a basic idea for the reuse of EoL components from passenger cars.

Above all, the unclear damage situation is a challenge that would have to be faced by damage analysis, including structural strength/stiffness analysis. Furthermore, the effort required for the necessary careful dismounting of the parts will be costly. Both would reduce the price advantage over a new component. This is a problem given the probably quite low customer acceptance for used parts, especially in the high-priced segment in which CFRPs are applied today. In addition, short design cycles limit the number or reuse cycles. In the automotive sector, this concept would probably only be of interest for the spare part market. However, this could change as development concerning damage analysis progresses, and other industries may be more suitable. In any case, if a reuse strategy is intended, this must be taken into account very early in product development, to ensure that damage-free dismounting is possible.

Recycling: This term describes the reconditioning of materials, products or substances for reuse for the original or a new purpose [66]. An example from the area of FRPs that is already industrially implemented today is given in the field of thermoset compression molding compound (SMC). Here, ground EoL-SMC is added to new SMC (particle recycling) [71]. Through this, the fibers again take over

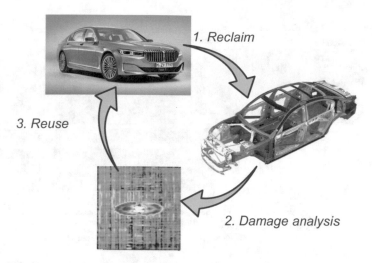

Fig. 5.26 Concept for the reuse of FRP components. Top and right image adapted, printed with permission of BMW AG, bottom image printed with permission of Leibniz-Institut für Verbundwerkstoffe GmbH

a load-bearing function. Thermoplastic molding compounds also allow component waste to be fed back into the process by simply melting them together with the new material. Repeated plasticizing can, however, affect the fiber length and fiber properties [72].

FRPs with continuous fiber reinforcement are somewhat more challenging with regard to recycling. Pyrolysis or solvolysis can be applied to separate fiber and polymer thermally or chemically, respectively. However, the fibers must then be treated with a sizing again in order to protect the fiber during further processing and to ensure proper fiber-matrix adhesion in the new part [73]. In the case of dry production waste, a further issue is to ensure purity of fiber type and finish. However, the most critical aspect of recycling with continuous fiber reinforcement is the fiber length preservation. Recovering the fibers causes a shortening and thereby a loss of orientation. If at all, fibers recovered this way are therefore only used for short or long fiber-reinforced FRPs, e.g. in SMC or as non-wovens [74, 75]. This is also referred to as "downcycling," since the original use in high-performance FRPs is not restored. A possible solution is offered by textile processes restoring the fiber orientation. In this process, the recycled, discontinuous fibers are spun into yarns and then processed to textiles [76]. These can then be further processed into FRPs. Figure 5.27 shows a process chain for the production of organo sheets from recycled carbon fibers. Compared to conventional textiles, corresponding materials can even offer additional functionalities, such as deep-drawing capability [76]. Corresponding developments are subject to research and have the potential to overcome the barriers to market entry.

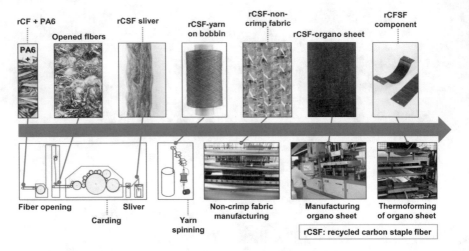

Fig. 5.27 Process chain for the manufacturing of organo sheets based on recycled carbon fibers. Adapted from [4] (Image adapted, printed with permission of Leibniz-Institut für Verbundwerkstoffe GmbH)

Both the processes for separating fiber and matrix as well as the processes for restoring fiber orientation are of less interest for glass fibers, as they quickly become economically inefficient, due to the comparatively low prices for new glass fibers.

Utilization/filling: These are processes in which the waste is put to a meaningful use, whereby meaningful means that the waste replaces other substances that would otherwise have been used for this purpose [66].

A classic example is the thermal recycling of waste in waste incineration plants. Due to the high calorific value (heat value) of plastics, this is also a viable option for FRPs. However, the fibers can cause problems, especially when they are highly temperature-resistant (e.g. carbon fibers) and impair filter systems. In addition, the much higher energy input for the production of matrix, fibers and FRPs, compared to the final calorific value, means that a high amount of energy is irretrievably lost [77].

Another option is to use FRP waste for cement production, which requires certain minerals in addition to large amounts of energy. GFRPs can first be thermally utilized, and the residues can then serve as raw material for the cement clinker. However, the main problem here is given by transport costs and professional fragmentation of large GFRP components such as wind turbine blades [78, 79].

Disposal: If none of the above-mentioned alternatives is possible, the last option is the disposal. According to the EU Directive, this includes all processes that do not constitute recovery. This does not necessarily mean that no substances or energy is recovered. Therefore, landfilling is a classic example. Even though the directive clearly defines disposal as the least preferred option, any higher-ranked recovery operation must ultimately be measured against this simple solution from a technical, economic and ecological perspective. Not every option that is technically possible is necessarily ecologically meaningful. All process steps toward recycling require resources themselves. In addition, an economically unfavorable recycling can reduce the resources for environmental protection elsewhere. Therefore, a holistic recycling strategy is necessary and starts with the component design.

In addition to the ones described above, there are of course also **other concepts** for the avoidance, reuse, recycling and utilization of FRPs. Furthermore, this area is receiving increasing attention in industry and research. In addition, when considering the challenges of recycling, it must be kept in mind that the use of FRPs is already economically and ecologically worthwhile. This is related to their specific advantages, such as lightweight potential, corrosion resistance and durability. Products made of FRPs therefore often show very positive impact on the life cycle assessment of products, even if reuse, as the ultimate goal besides waste avoidance, cannot yet be achieved everywhere. However, for all approaches one can state that a meaningful strategy for a closed material cycle requires not only the development of appropriate processes, but also holistic, integrated product development. This also includes an appropriate design of components (e.g. design with recycled materials). In the following section, it will be shown how this can be achieved.

For design to recycle, generally the same guidelines apply as to other materials. Pahl and Beitz [80] define the following guidelines that are relevant for a complete product or assemblies. They can be applied both individually and in combination:

1. **Material compatibility**: Since recycling-friendly single-material products are rarely meaningful, inseparable combinations of materials should contain materials that are compatible during recycling so that they can be recycled economically and with high quality [80]. → This requirement points out the basic problem of FRPs when it comes to recycling. Polymers and fibers can only be considered as compatible, if they can be further processed together, as it is, e.g., the case for thermoplastic FRP components recycled into a molding compound. Otherwise, a complex separation is required. The only real exception is given by self-reinforced thermoplastics, in which fiber and matrix are of the same basic material [81]. In addition, beyond material level, compatibility of FRPs and other materials joined to them, e.g. metals, is mostly not given. This means that separation is required for recycling. Nevertheless, material compatibility is an important topic for FRPs, e.g. if they are joined with pure polymers or when FRP components are back-molded in an injection molding process. Here, the use of compatible polymers is important.

2. **Material separation**: If material compatibility cannot be achieved for inseparable parts and groups of a product, then additional joints are to be incorporated in order to allow separation during recycling, e.g. by disassembly [80]. → Easily separable joints, e.g. screw joints, are a challenge for FRPs, due to the induced notch effect. They often require additional local reinforcement, which counteracts lightweight intentions (see Sect. 5.3.2).

3. **Reprocessing-compatible joints**: Joints that are suitable for qualitative and economical recycling should be easy to dismantle and to access and should

possibly be arranged at the outer product zones. Multi material design generally requires a higher recycling effort and should be avoided [80]. $\rightarrow$ Easily dismantled joints often include precisely those that are not regarded as suitable for FRPs, e.g. screwed connections. Accordingly, design to recycle is today often neglected especially concerning this point.

4. For **economic disassembly** the use of simple tools, automated systems and/or untrained workers should be enabled [80]. $\rightarrow$ If the separation requires the mechanical processing of the FRPs, e.g. by sawing, containment is usually necessary to protect humans and the environment from the generated dust. This is especially true for CFRPs, where recycling is of main interest. For the mechanical processing, special tools are often used and appropriate training of the employees with regard to the risks is essential. This requirement can therefore only be addressed indirectly, by selecting the joining process in a way that the FRP components themselves do not need to be mechanically processed.

5. **High-quality materials**: Valuable and rare materials particularly need to be arranged and labeled in a manner suitable for dismantling [80]. $\rightarrow$ Even though CFRP is an expensive material in the production process, it will only be regarded as valuable waste, when high prices can be received for scrap material. This in turn requires industrially established recycling strategies. With ongoing further development of appropriate technologies, CFRPs can then indeed become a valuable material for professional recycling companies. In this context, purity of fiber type and possibilities for identification of the fiber type will play an important role.

6. **Hazardous substances**: Substances which pose a hazard to humans, systems and the environment during recycling or reuse must always be arranged in a separable or drainable manner [80]. $\rightarrow$ It is known that FRPs can lead to problems, e.g. in thermal recycling if the fibers do not burn without leaving residues and thus clog filter systems [77]. Accordingly, a separation from other materials is necessary and must be addressed accordingly in the design.

Regarding these guidelines for a design to recycle, one must keep in mind that these exclusively serve this single objective. The effect of according measures on other targets of the product development therefore needs to be considered. This is especially challenging for FRPs. Quickly separable connections, e.g. screwings, are often not FRP-suitable and lead to weight increase, which counteracts the frequently pursued goal of a lightweight design. Balancing the different goals is a classic task of the IPD.

In **summary**, the following can be said about the design elaboration:

- An anisotropic design offers several opportunities, but also risks.
- Manufacturing requirements for the design on the one hand and production-related influences on the component performance on the other hand must be taken into account.
- Many problems, especially in the context of force application, joining, repair and recycling must be solved differently compared to other material groups.

- A high technical value of a solution often leads to decreasing cost-efficiency, and therefore a holistic consideration is necessary.

5.4 Elaboration of the Process Concept

The elaboration of the process concept includes the selection of suitable production systems and the process design. These steps are further discussed in the following section.

5.4.1 Selection of Production Systems

Taking into account the selected semi-finished products as well as the targeted boundary conditions concerning cycle time, cost-efficiency and component quality, the requirements for the production systems can be derived. The systems required for the production of FRPs differ from process to process. Table 5.13 lists typically used production systems for the different processes and process groups, respectively. Note that the individual processes of a process group can differ considerably and the used semi-finished product may strongly influence the necessary facilities. The list therefore only provides a basic overview. Furthermore, the systems required for the production of the semi-finished products are generally neglected, despite the fact that "make or buy" decisions, i.e. decisions on whether to buy the semi-finished product or manufacture it in-house, depend on many factors. External procurement does not necessarily need to be the cost-efficient option. As Table 5.13 shows, the spectrum of systems used to produce FRPs is broad. An individual consideration of the specific requirements for each single system is therefore not meaningful here. However, to give a first insight, Table 5.14 contains a checklist of typical requirements that must be taken into account, when planning the process infrastructure. When procuring production systems, these criteria must be taken into account keeping in mind the specific application. Therefore, it should always be checked if the production capacity of the systems is sufficiently utilized or will be in the future. If not, it should be ensured that the system can also be used to manufacture other products. Cost and performance need to be balanced. If there are too many unknown factors, different systems must be considered as alternatives in the economic evaluation (Sect. 6.2). For this, however, an estimation of the cycle times achievable by the different systems is required (see Sect. 5.4.2.1).

Table 5.13 Commonly used production facilities depending on the manufacturing process

Process	Production technology
Fiber spraying	TS[a]: fiber spray gun, tool, spray cabin, possibly oven
Centrifugal casting	TS: mold, rotation system, injection system
Winding	TS: spool stand, resin impregnation unit, winding system, core tool/liner, oven with rotation system TP[b]: spool stand, heating system, winding system, core tool
Pultrusion	TP: spool stand, heating system, form tool with cooling unit, pull-off device, cutting unit (e.g. flying saw) TS: spool stand, impregnation unit + forming tool or injection tool, pull-off device, cutting unit (e.g. flying saw)
Extrusion	TP: extruder, heated extrusion tool, possibly a cooling unit, cutting unit (e.g. flying saw)
3D printing (continuous reinforcement)	TP: 3D printer
Tape laying	TP: spool stand, industrial robot/gantry system, tape laying head, table and tool for lay-up
Autoclave	TS: lamination room with venting, eventually automated lay-up system (e.g. tape layer), cooling systems for prepreg, tool, autoclave
Prepreg compression molding	TS: lamination room with venting, eventually automated lay-up system (e.g. tape layer), cooling systems for prepreg, heated tool, press
Thermoforming	Heating unit (e.g. infrared field or oven), transfer system (e.g. frame) for transfer of molten organo sheet to press, heated tool, press
Hand lay-up	Tool, lamination room with venting
Liquid composite molding	Eventually CNC cutter and preforming tool, heated injection system, tool carrier
Compression molding	TP: plasticizer, heated tool, press TS: cooling unit for prepreg, heated tool, press
Injection molding	TP: plasticizer, heated tool, closing unit
Post-treatment	Trimming system (e.g. water jet or laser cutter or milling machine)
Joining technology	Industrial robot/gantry system with application unit for welding, riveting or adhesive bonding
Transport between process steps	Industrial robot with handling system

[a]TS = Thermoset
[b]TP = Thermoplastic

5.4.2 Process Design

During process design, the interaction of the selected production systems is optimized. Thereby the focus, besides achieving the required quality, is to minimize the cycle time. This requires to take the following steps:

Table 5.14 Examples of basic requirements for production systems

Requirements		Details, examples
	Processable part dimensions	Minimum and maximum component area that can be processed economically. The shape (in particular length and width) must be taken into account. The processable component dimensions often directly correlate with the tools that can be mounted in a system
	Manufacturable part depth	Has direct influence on the height of the tool, which must be mountable. When pressing, the maximum depth is defined by the maximum press stroke
	Press force, locking force	Depending on the process, pressures emerge in the mold, which must be counteracted in order to close the mold or keep it closed. For example, injection pressures during injection molding or forming forces during thermoforming
	Moving velocities	Many systems have elements that have to move in the course of the process, whereby the speed has a direct effect on the cycle time. For example in tape laying machines, presses or rotating winding cores
	Temperature ranges, heating and cooling velocities	The polymer materials to be processed define the temperature control requirements. Therefore, the system parameters should be adjusted accordingly. The heating and cooling speeds have a direct effect on the cycle time
	Volume flow rates and pressure ranges for material conveyance	Injection systems, extruders, plasticizers, etc., must cover certain volume flow rates and pressure ranges, depending on the component volume and the semi-finished product properties, in order to enable (economical) production. Since a high maximum value often also defines the minimum value, exact planning is required here
	Heating performance	Heating equipment is required for temperature control, for example, when heating semi-finished products such as organo sheets. Their performance has an

(continued)

Table 5.14 (continued)

Requirements		Details, examples
		effect on energy consumption and the achievable cycle times
	Installation space	Installing a system requires space and thus leads to space utilization costs that depend on the size of the system. Different design methods can result in significant differences (e.g. press with hydraulic cylinders next to or on the press)
$	Investment costs	The costs and performance of a system must be balanced
	Ergonomics/operation	Easy operability of the system software reduces the training effort for employees and reduces the potential for errors. An ergonomically favorable mode of operation potentially reduces costs due to sickness of staff

1. Determination of the processing characteristics of the semi-finished products.
2. Estimation of the cycle time for all individual steps and optimization of the configuration of the production facilities.
3. Optimization of the number of production systems for series production.
4. Sensitivity analysis.

The determination of the relevant semi-finished product characteristics has possibly already been carried out during the selection of semi-finished products (see Sect. 5.2.1.4). The following steps are explained below.

5.4.2.1   Cycle Time Estimation and Optimization of the Configuration of Production Systems

Several process steps typically occur during the production of FRPs and must be evaluated concerning the cycle time. In the following section, it will be outlined how this can be done for the different process steps.

Heat transfer

Heat transfer plays an outstanding role in the production of FRPs. All thermoplastic prepregs must be heated before processing, so that the thermoplastic matrix can flow or be deformed. For solidification, the heat must be extracted again. For thermoset FRPs, heated tools are often used to influence the viscosity behavior of the resin system and to accelerate the curing reaction. The cycle time for heating/ cooling is affected by the production systems and the semi-finished products. Concerning the latter, specific restrictions must be considered, such as the influence of the resin temperature on the maximum possible processing time and the viscosity. The calculation of the cycle times is a classical question of heat transfer theory, whereby different heat transfer mechanisms must be taken into account, depending on the production technology.

Conduction refers to heat conduction within a continuum. This is the dominant mechanism, when a FRP is heated or cooled by direct contact. Figure 5.28 shows an example of the simplified case of a flat plate that is heated by one-sided contact with a mold. It is assumed that the plate is thermally perfectly insulated on the sides that are not in contact with the mold. In this case, the result is a one-dimensional, stationary heat conduction. Due to their structure, one-dimensional heat conduction in FRPs is often a very strong simplification. It should therefore only be used for a first estimation.

If the upper side of the FRP plate is brought into contact with a tool of the temperature T_1, while the FRP plate consistently has the temperature T_2, with $T_1 > T_2$, then the result will be a temperature gradient over the thickness of the plate. The heat energy transferred per time is called heat flow ($\dot{Q}$) and can be calculated for simple geometries as follows [82]:

$$\dot{Q} = \int \dot{q}\, dA \tag{5.2}$$

With

$$\dot{q} = -\lambda_w \frac{dT}{dx} \tag{5.3}$$

Here $\dot{q}$ is the heat flux density perpendicular to the part surface A, which is proportional to the thermal conductivity λ_w. The thermal conductivity describes the transferred heat flow per length unit dx and per Kelvin temperature difference. If the thermal conductivity is constant (a proper assumption for solids and liquids where the influence of temperature and pressure is quite small [83]), the heat flow over the thickness of a plate with the thickness s can be calculated as follows [82]:

$$\dot{Q} = -\lambda_w A \frac{dT}{dx} = \frac{\lambda_w}{s} A (T_1 - T_2) \tag{5.4}$$

The thermal conductivity of the FRP can be determined, for example, by a simple mixing rule based on the individual components and their volume proportions [84]. A more precise result is obtained by micromodeling of the FRP and numerical determination of the thermal conductivity using appropriate software [85]. In both cases the thermal conductivity of typical polymers and fiber materials, as individual materials, can be taken from the relevant literature (e.g. [86]).

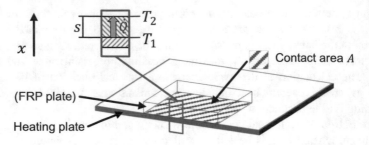

Fig. 5.28 One-dimensional, stationary heat transfer in a (FRP) plate heated by tool contact. Adapted from [82]

Another typical case is **convection**, i.e. cooling or heating by heat transfer between a fluid and a solid. This occurs, for example, when a FRP is processed in a conventional convection oven or in an autoclave. A distinction must be made between free convection and forced convection. In free convection, a fluid at rest meets a solid surface with a different temperature, which leads to differences in density and corresponding compensating flows. In forced convection the fluid flow is maintained by an external pressure difference [87]. The heat $\dot{Q}$ absorbed or emitted from a body with the area A and the temperature T_K can be calculated as [68, 88]:

$$\dot{Q} = \alpha A(T_K - T_F) \tag{5.5}$$

Here T_F is the ambient temperature and α is the convective heat transfer coefficient. The heat transfer coefficient in turn depends on numerous boundary conditions, such as the flow conditions (Reynolds number), the component geometry, the material properties and the direction of heat transport. For cases that allow a strong simplification, such as the heating of a square plate (100×100 mm^2) by free convection (see Fig. 5.29), however, a first estimation using an analytical solution based on the so-called Nusselt number Nu is possible. This is shown in the following based on [82, 87, 88].

First, one can define

$$\alpha = \frac{Nu \cdot \lambda_W}{L} \tag{5.6}$$

L, the characteristic length, can be calculated from the edge length z of the FRP plate.

$$L = \frac{z^2}{4z} \tag{5.7}$$

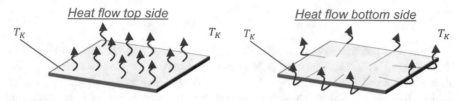

Fig. 5.29 Free convection on a flat plate. Adapted from [82]

For the top and bottom side Nu results from the corresponding Prandtl function for the top side ($f_{\text{top side}}(Pr)$) and the bottom side ($f_{\text{bottom side}}(Pr)$) as well as the Rayleigh number Ra, with

$$f_{\text{topside}}(Pr) = \left[1 + \left(\frac{0,492}{Pr}\right)^{\frac{9}{16}}\right]^{-\frac{16}{9}} \approx 0.3490 \tag{5.8}$$

and

$$f_{(\text{bottom side})}(Pr) = [1 + (0,322/Pr)^{(11/20)}]^{(-20/11)} \approx 0.4055 \tag{5.9}$$

$$Ra = Gr \cdot Pr \tag{5.10}$$

The difference between top and bottom results from the fact that the air, which is reduced in density by heating, must flow under the plate to the edges in order to be able to rise. At the top, it can rise directly. Pr is the Prandtl number which can be assumed to be 0.719 for air at 500 °C, resulting in the given values. The Grashof number Gr is derived from a series of thermophysical quantities of air and the acceleration due to gravity g.

$$Gr = \frac{g \cdot L^3 \cdot \beta \cdot (T_K - T_F)}{v} \tag{5.11}$$

The values can be assumed as follows: β as the coefficient of thermal expansion for an ideal gas with 0.00129341 1/K and v as the kinematic viscosity of air at 500 °C with 81.4E-6 m^2/s. Eventually, α and thus the heat flow for top and bottom side can be calculated.

Finally, the third standard case is **heat radiation**, i.e. heat transfer by electromagnetic waves. This occurs, for example, when organo sheets are heated by an infrared radiant heater. The heat flow $\dot{Q}_{1 \to 2}$ transferred by radiation from a radiator of temperature T_1 to a FRP panel with area A_1 and temperature T_2 can be calculated as follows, according to [82]:

$$\dot{Q}_{1\rightarrow 2} = F_{1\rightarrow 2} \cdot \sigma \cdot A_1 \cdot \left(T_1^4 - T_2^4\right) \tag{5.12}$$

Here σ is the Stefan-Boltzmann constant $(5.670*10^{-8}\ \mathrm{W/(m^2K^4)})$ and $F_{1\rightarrow 2}$ is the so-called visibility factor describing the position of the plate in relation to the radiator. It can be calculated by taking into account the angles β_1 and β_2 under which the surfaces appear to each other and the distance H between the radiation source and receiver.

$$F_{1\rightarrow 2} = \frac{1}{\pi A_1} \int\limits_{A_1} \int\limits_{A_1} \frac{\cos\beta_1 \cos\beta_2}{H^2} dA_1 A_2 \tag{5.13}$$

Often combinations of conduction, convection and radiation occur. In addition, depending on the polymer, endo- or exothermic reactions have to be taken into account.

Impregnation

Manufacturing FRPs requires the combination of the fiber structure and the matrix polymer. For endless fiber-reinforced polymers, this means that the fiber structure must be impregnated with the matrix polymer in liquid form. The starting point for the description of the impregnation process is Darcy's law, which has already been presented in the context of the characterization of the permeability of fiber structures (Eq. 5.1). Generally speaking, Darcy's law can also be used to estimate the cycle time of impregnation processes. However, contrary to permeability measurement, it is not possible to adapt the process to the validity conditions of the law. Instead, it must be checked whether the given process boundary conditions allow the application of the law. According to Neumann [89], Darcy's law can be derived from the Navier-Stokes equation, when the nonlinear inertial elements in this equation of motion are omitted. In order for these simplifications to be valid and for the law to be applicable to unsaturated flow, a number of boundary conditions must be met:

- Darcy's law describes a continuous, saturated flow. The impregnation of a fiber structure is an unsaturated, progressive process in which an initially dry fiber structure is gradually impregnated. The equation can be adapted to this by substituting the quotient of volume flow and cross-sectional area by the flow velocity $\vec{v}$, as shown in Eq. 5.14 for a one-dimensional flow. $\vec{v}$ is the so-called volume-averaged flow velocity (the actual flow front velocity is the quotient of volume-averaged flow velocity and porosity). p corresponds to the pressure gradient, i.e. the pressure drop between inlet and flow front related to the corresponding flow length. Depending on the process, ambient pressure or vacuum is applied at the flow front and the injection pressure is a process parameter to be set. [8]

$$\vec{v} = -\frac{K \cdot \nabla p}{\eta} \tag{5.14}$$

For this formula to be at least approximately applicable, further conditions must be fulfilled, which are considered in the following.

- The **flow** must be **laminar** (not turbulent): If the flow is not laminar, there is a nonlinear correlation between injection pressure and flow velocity. This is not represented by Darcy's law. To validate applicability, the Reynolds number Re, which is the ratio of inertial forces and frictional forces [90], must be checked. Below a critical value, the flow can be assumed to be laminar [91]. This critical value is usually given as 1 [92] or 1–10 [93]. For resin injection processes Eq. (5.15) can be used for calculation [94]:

$$Re = \frac{v\rho\sqrt{K}}{(1 - V_F)\eta} \tag{5.15}$$

Here v is the flow velocity, ρ is the density, η is the fluid viscosity, K is the textile permeability and V_F is the fiber volume content. For a typical RTM process, these values are in the range of 0.1 [95], so the applicability is justified.
- There are **no changes in the cross section** through which the flow passes (mechanically stiff porous media): This assumption is usually considered to be sufficiently fulfilled for a primarily in-plane (textile plane)-oriented flow front propagation [96]. In the case of flow front propagation perpendicular to the layer structure, however, strong deformations can occur due to hydrodynamic textile compaction, induced by the pressure conditions generated by the flow [97]. The high melt viscosity during impregnation with thermoplastics can also lead to considerable textile deformation [98], which contradicts application of Darcy's law.
- There is **no influence of capillary forces**: The capillary pressure present in technical textiles depends on the textile architecture, the direction of flow, the fiber volume content and the fluid, whereas the in-plane capillary pressure is approximately twice as high as the out-of-plane capillary pressure, at equal fiber volume content and fluid [99]. In his studies with epoxy resin, Ahn was able to determine capillary pressures up to 0.37 bar [99]. Studies by Gibson, also with thermoset resin systems, showed capillary pressures that were about an order of magnitude smaller [100]. This shows that capillary pressure of technical textiles can strongly vary. Despite all this, capillary pressure can be neglected for most injection-based liquid composite molding processes, such as RTM, due to the significantly higher injection pressures compared to the capillary pressure. However, for infusion processes, where the pressure gradient is naturally less than 1 bar (vacuum vs. ambient pressure) the capillary pressure can have quite an influence.
- The flowing fluid must be **Newtonian** (no shear rate dependence) and **incompressible**: The assumption of a shear rate-independent viscosity is widely meaningful for thermoset resin systems. The situation is different, however, with thermoplastics. Here, with increasing shear rate, the molecules are aligned in the direction of flow, which reduces the internal friction and thus the viscosity [101]. The description of a textile impregnation by a thermoplastic, using Darcy's law, is therefore not easily possible. Mayer [90] developed a model for the transverse

impregnation of textiles with thermoplastics, as it takes place in the production of organo sheets. In order to compensate for the thickness changes caused by flow-induced deformation as well as for the viscosity changes the so-called B-factor is determined, based on an organo sheet production process that has led to full impregnation. This factor quantifies the impregnation performance of a process with regard to the influencing variables time, temperature and pressure. Thus, the need for explicit modeling is bypassed by empirically determined parameters. Christmann [102] extended this model to the case of varying pressure conditions.

Even though there are some influences that cause deviations between prediction and reality, Darcy's law offers an efficient way of estimating cycle time, especially for resin injection processes. Therefore, it is nowadays the basis of all common simulation programs for the numerical description of liquid composite molding processes, such as PAM-RTM [103], SimLCM [104], RTM-Worx [105] or LIMS [106]. The input data fed to these programs comprise the process boundary conditions, the orientation- and fiber volume content-dependent permeability as well as the temperature and time-dependent viscosity of the resin. Methods for the characterization of these properties have already been explained in Sect. 5.2.1.4.

Transport Procedures

Between the individual process steps, the semi-finished product must be transported, which includes picking it up and positioning it in the next system. The required times can usually be estimated based on experience, since the share of the total cycle time is comparatively low and inaccuracies are therefore less significant. For very time-critical applications (large series), however, a detailed consideration can take place. In the case of an automated process, this requires details on the system characteristics, e.g. travel speeds and acceleration times of an industrial robot or a gantry system. However, it should be taken into account that the maximum performance of the robot cannot necessarily be exploited, for example, because accelerations cause the transported goods to slip or can induce uncontrollable oscillation of the gripper system.

Kinematics

A significant part of the cycle time of a process results from simple mechanical movements that can be calculated by the corresponding, common formulas. This, e.g., applies to the **forming process** for an organo sheet. The forming process consists of the closing and opening process as well as the holding phase, whereas the latter is determined by the time necessary to cool down the part below melting temperature. The times required for closing and opening can be derived from the travel speed of the selected press, as well as the opening height required for part positioning and the forming depth (resulting from the depth of the component). For the closing phase up to the initial contact with the semi-finished product, the full potential of the system can be exploited. However, for the actual forming process mostly the forming behavior of the organo sheet determines the limit and not the technical capability of the press. An estimation of a reasonable travel speed must be based on experience. Once all the values have been identified, cycle time for the single movements can be calculated quite easily, taking into account acceleration and deceleration phases. The calculation of the holding/cooling time is done by applying the heat transfer formulas presented earlier, taking into account the heating power and the limitations induced by the material.

Another example is post-treatment by edge trimming: The cycle time for this process can be calculated from the component geometry (length to be trimmed) and a reasonable trimming speed. Further typical kinematic processes can be found at all processes where an industrial robot or a gantry system is used (e.g. tape laying). For these processes, it is often possible to fall back on preexisting studies, allowing detailed cycle time estimation, for example, in the form of a collection of formulas as developed by Beresheim for tape laying [107] or in the form of a software like CompositeCAD® or CADWIND® for winding processes.

Mass- or volume-related simplifications

The effort for cycle time estimation can be reduced significantly, when applying time factors related to mass or volume. For example, for tape laying an average placed tape mass per hour can be used for a first estimation. A component-specific cycle time estimation would require considering all acceleration processes, at the beginning and the end of each placement path and at radii. It is therefore very complex and work-intensive. With this in mind, the error induced by the mass-related simplification often seems acceptable. However, it should be kept in mind that corresponding values stated by manufacturers of production systems naturally are quite optimistic values and can only be achieved under idealized conditions.

The formulas previously presented intend to show the reader that a profound cycle time estimation is already possible with comparatively simple measures. If, at this point, there are still various alternative production strategies available for selection, this estimation can be used directly, if necessary, to evaluate the economic advantages of the alternatives in the next step (Sect. 6) and to boil down the number of alternatives. This way, the effort for detailed numerical modeling for cycle time estimation can be reduced. Furthermore, the "manual" cycle time estimation fosters an understanding of the process physics. This understanding is necessary to develop a feeling for how decisions—from all areas of IPD—affect the process. Therefore, it is also useful for the IPD team members who are not directly involved in the process design to understand these basic formulas. More detailed explanations on the modeling of FRP manufacturing processes can be found, e.g., in [108].

Numerical process simulation for detailed cycle time estimation

Even if the estimation based on the basic formulas often already gives valuable insights, numerical simulation is the method of choice for an accurate prediction and an efficient optimization. A process simulation allows pursuing different goals:

- Detailed study of the process functionality (interaction of parameters).
- Efficient parameter variation for optimization (cycle time, robustness, etc.).
- Support of the tool design
- Identification of process-related optimization potential for the component design (design to manufacture).
- Identification of possible risks in terms of manufacturing defects.

For the numerical simulation of FRP manufacturing processes, classical multi-physics software packages such as Ansys or LS-DYNA are used. In addition, some specialized software packages are available commercially or as "open source."

At this point it should be noted that the simulation programs available for FRPs are often comparatively new, compared to solutions for other material groups, and hence not always to be considered fully mature. Nevertheless, the developments in this area have advanced rapidly, and today, numerical simulation is an indispensable support for IPD. Through numerical simulation, numerous alternative process boundary conditions can be evaluated concerning their influence on cycle time and robustness, with manageable effort. However, detailed approaches to numerical simulation are not covered by this book.

5.4.2.2   Optimization of the Number of Production Systems for Serial Production

The previous considerations allow the calculation of the cycle time for all single process steps involved. In serial production, they can possibly be parallelized. In turn, this means that individual process steps can form a "bottleneck" causing an economically unfavorable low capacity utilization of the systems used for other process steps. If multiple systems are used for this "bottleneck" process step, the capacity utilization can be improved over the entire process chain, which may lead to an overall improved cost-efficiency due to decreasing fixed costs. To illustrate this, Fig. 5.30 shows an example. The considered process chain comprises the process steps preforming (120 s), resin injection (240 s) and eventually a post-treatment (edge trimming, 60 s). Before and after the single process steps, transportation (10 s) takes place. Figure 5.30 first shows the case of single-part production. Here the steps take place one after the other, resulting in a total cycle time of 450 s. However, the process steps are carried out on independent systems and can take place simultaneously when it comes to serial production. Of course, from the point of view of a single component, the processing time remains the same, but the effective total cycle time for parallelized production is significantly reduced to 250 s. The effective cycle time is relevant for the calculation of the annual output, as it defines the time between the completions of two components. Closer examination reveals that the injection process represents the "bottleneck," causing the preforming line capacity to be only incompletely utilized. Therefore, to achieve higher quantities an additional set of systems for the injection process can be put into operation. Through this, the effective cycle time can be further reduced to 140 s.

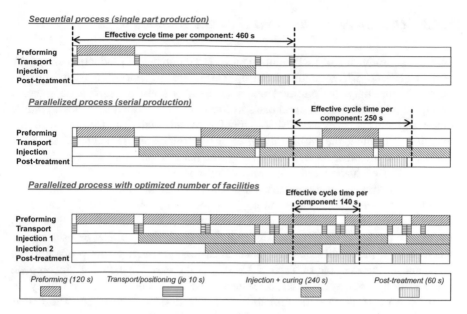

Fig. 5.30 Influence of parallelization and optimization of the number of systems on the effective cycle time

5.4.2.3   Sensitivity Analysis

Using the mathematical or numerical modeling of the cycle time, a detailed sensitivity analysis should be carried out, in order to identify potential for cost savings. The main aim is to answer the following questions:

- What are the most **time-critical process steps**?
- How large are the potential savings through **automated handling technologies**?
- Which **system characteristics** have the greatest influence on the cycle time and how much effort is required to modify them?
- How great is the potential for savings through alternative, **more efficient systems**?
- Which **geometric component features** have the greatest influence on the cycle time and can they be modified?

If large potentials for savings are identified, it must be checked whether automation, modification of the system technology or even adjustments to the component design should be carried out. Since at this point only the cycle time is calculated, but not the final production costs, no final statement can be made on the actual economic effects of such changes. However, since cycle time and manufacturing costs are usually strongly correlated, large time saving potentials are a quite good indicator. A detailed consideration of an alternative with high time saving potential, in terms of an economic evaluation (see Sect. 6.2), can therefore be worthwhile.

5.4.3 Quality Assurance and Damage Analysis

During the production of FRP components a large number of production errors can occur (see Sect. 5.3.1), ranging from optical defects to serious errors that impair safety. Accordingly, the question arises as to how these errors can be detected during quality assurance. A good basis in this context is formed by a classic failure mode and effects analysis (FMEA). FMEA is an analytical method with the intention to evaluate the probability of error occurrence on the one hand and error discovery on the other hand. Additionally, the damage potential can be taken into account. The evaluation can, e.g., be carried out on scales from 1 to 10, where defined descriptions or frequencies are assigned to the numerical values. An introduction to the FMEA methodology is provided, e.g., by [109]. Table 5.15 shows extracts from an FMEA for a component with unidirectional fiber reinforcement, produced via resin transfer molding. Particularly critical cases are such, where a high probability of occurrence goes together with a low probability of detection. In the example of Table 5.15, this is the case for the error "incomplete impregnation." For such cases, measures either to reduce the probability of occurrence or to increase the probability of detection must be taken. The first is preferable and can be achieved through appropriate adjustments in the design and the manufacturing process.

An **increase in the probability of detection** can be achieved by applying suitable **technologies for nondestructive damage analysis**, which play a special role for FRPs. A FRP is not a material that is first produced in a standardized mass process and then further processed into a component, while maintaining its basic material structure. The FRP and thus its final properties as a material are only created during the actual component production. That is why damage analysis must be done on the often very complex part geometry. Methods of nondestructive testing are mostly based on the evaluation of properties such as heat or sound propagation in the component. Due to their internal structure, these properties are direction-dependent for FRPs and strongly weakened by damping, scattering and reflection at the extremely numerous internal interfaces. Furthermore, the errors to be detected are often in the same size as the structural elements of the FRPs, such as roving diameter or fiber layer thickness. The new and further development of corresponding methods therefore remains an important research topic. Some of the most important methods are briefly presented below.

Knock test: This very simple test method is still the most important one. Here the component is fixed and knocked on using a small hammer or a coin. The sound resulting from the local excitation is evaluated by an examiner. In the case of a local defect, such as delamination, a rather damped sound can be heard compared to an intact area. This is because the defect influences the damping of the knocking impact. Of course, the method depends heavily on the subjective experience of the examiner; however, it is generally still possible to distinguish between different defect types and to estimate their extent. In order to reduce subjectivity and thereby improve reliability and reproducibility of the method, today there are a number of

Table 5.15 Example for a FMEA for the manufacturing of an FRP part via resin transfer molding (excerpt)

Defect	Probability of appearance	Potential risk	Probability of detection
Incomplete impregnation of unidirectional reinforcement	8—moderate (1:50) →very low permeability in UD-reinforced areas	10—very serious → safety may be impaired	10— < 90% → pure visual inspection not sufficient
(Relative) fiber displacement through injection	3—very low (1:2000) → use of cross-linkable binders minimizes the probability of appearance	9—very serious → safety may be impaired	6— > 98% → detectable by automated 100% inspection, since injection starts from cover layer and displacement is greatest there
Surface defects	8—moderate (1:50) →induced by irregularities in preform	4—fairly serious defect → leads to customer dissatisfaction but no safety impairment	2— > 99.7% → obvious error feature, detectable by automated 100% inspection
...	...	...	...

automated, electronic measuring aids available, which are comparably easy to apply and allow data acquisition [110, 111].

Ultrasonic testing: A well-known biological equivalent of this method is the echolocation of animals like bats, which can orient themselves even in complete darkness by application of the pulse-echo principle (see Fig. 5.31, left). To test a FRP component, a sound wave in the part is generated by a probe. This wave is introduced into the FRP component via a coupling medium (typically water). In the usually applied pulse-echo method, the sound waves travel through the component, and any reflections are then again absorbed by the probe. The resulting echoes are resolved over the transmission time of the ultrasound signal and interpreted electronically. The acoustic behavior of the defective area deviates from non-defective areas, if the error leads to reflection, scattering or diffraction. Thus, these are detected (see Fig. 5.31).

Generally, three types of ultrasound scans can be distinguished [112]:

- A-scan: local, punctual scan at the probe position.
- B-scan: probe is moved along one axis, resulting in multiple A-scans, which provide a cross section over the thickness.
- C-scan: probe is moved along two axes, so that a flat sectional image is produced at a defined depth or over the entire thickness.

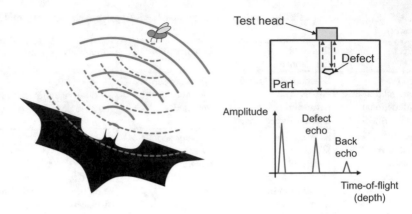

Fig. 5.31 Schematic illustration of the echo localization of bats (left) and ultrasonic scanning of a FRP component (right)

Figure 5.32 shows an example of a CFRP part (top) that was manufactured with purposely induced errors. It contains drill holes with different depths and varying diameters, as well as FEP[1] foils of different sizes in different depths. The latter simulate a delamination, which leads to sound reflections due to the additional boundary surfaces. The picture below shows the corresponding C-scan. All defects are very clearly visible in the ultrasound image, which is why they are classified as so-called visible defects. Today, the ultrasound method is a standard procedure for the quality assurance of FRP parts in aeronautics. However, the extensive scanning process makes it very costly.

Shearography: This term is the short form of laser speckle shearing interferometry, an optical method for nondestructive inspection of components. Here, a component is illuminated with an expanded laser beam. The surface of the component reflects the laser beam, which is then passed through a shear element and into a digital camera. The shear element creates two slightly shifted images of the object, the so-called interferogram. If the component is subjected to a load, e.g. thermal (e.g. halogen spotlight), mechanical (e.g. vacuum hood) or dynamic (piezo-shaker), intralaminar defects become apparent by a bulge on the surface. The bulge locally causes an additional phase shift in the interferogram (see Fig. 5.33). Correlation of interferograms of different load conditions results in a phase difference image, in which the differences become visible, so that conclusions can be drawn about material inhomogeneities under the component surface [113, 114]. Figure 5.34 shows a phase difference image for the already presented reference part. A vacuum hood was used as loading method. The figure shows that the drill holes are not visible and are to be classified as "non-visible defect." This is because vacuuming has no effect on the holes. Depending on their size and depth, the film inclusions are barely visible ("barely visible defect") or not visible.

---

[1]Fluorinated ethylene propylene.

Sample plate

Dimensions: 280 x 230 mm, thickness 1.45 mm

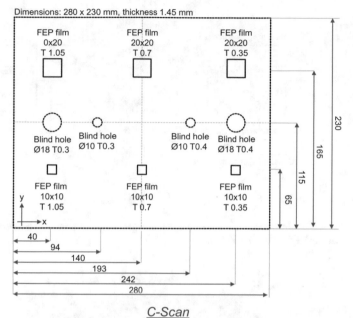

C-Scan
Amplitude of back side

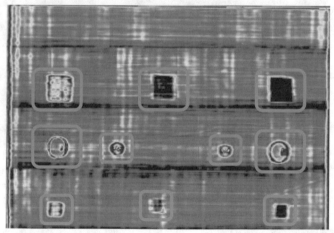

Material: CF-EP (T800SC 24K / M21) with layup $(0°/90°)_{4s}$

Legend: [VD]   [BVD]   (NVD)   VD —visible defect
BVD —barely visible defect
NVD —non-visible defect

Fig. 5.32 Ultrasonic scanning of a reference part with implemented defects. Images adapted, printed with permission of Leibniz-Institut für Verbundwerkstoffe GmbH

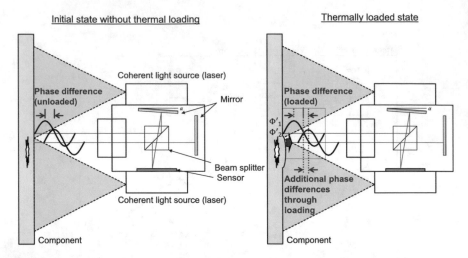

Fig. 5.33 Schematic illustration of a shearography. Images adapted, printed with permission of Leibniz-Institut für Verbundwerkstoffe GmbH

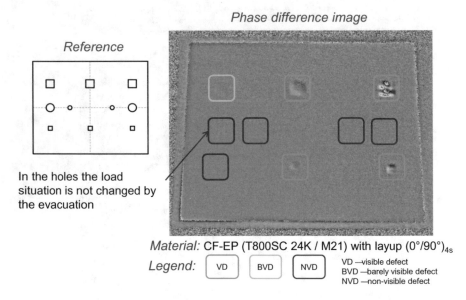

Fig. 5.34 Shearography of a reference plate with implemented defects. Images adapted, printed with permission of Leibniz-Institut für Verbundwerkstoffe GmbH

Thermography: Every body emits energy in the form of electromagnetic radiation. Electromagnetic radiation hitting a body is partly absorbed, partly reflected and partly transmitted. By means of thermography, defects can be detected by their influence on the thermal conductivity and the heat capacity of a material.

For this purpose, the surface is thermally loaded, for example, by an energy-intensive flash of light, and the surface temperature is observed. Here, for FRPs, the low thermal conductivity and the anisotropy are particularly challenging [113–116]. Figure 5.35 shows the scheme of a corresponding test setup and an example component with resin-rich areas. These become visible due to the different reactions. Table 5.16 lists the advantages and disadvantages of the three methods.

Of course, there are also other technologies for damage analysis. Figure 5.36 shows a comparison of different methods with regard to the detectable material defects.

Without claiming to be complete, these methods represent an extract of today's possibilities in the field of nondestructive testing of FRPs. It is important to note that there is no generally applicable method. A specific consideration of possible errors is necessary to find the appropriate method. Combinations of different methods are often useful to ensure a comprehensive examination.

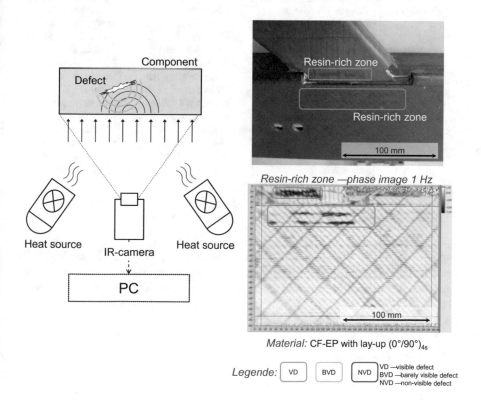

Fig. 5.35 Thermography of a CFRP component with defects. Images adapted, printed with permission of Leibniz-Institut für Verbundwerkstoffe GmbH

Table 5.16 Advantages and disadvantages of ultrasonic, thermographic and shearographic tests

	Ultrasonic	Thermography	Shearography
Advantages	+ Information about type and depth of error + High lateral resolution + High depth range and resolution	+ Optical test procedure—contactless + Simple and mobile measurement setup + Areal testing + Short measuring time + Viewing angle can vary + Relative measuring method (robust against disturbing influences) + Defect selective	+ Optical test procedure—contactless + Simple and mobile measurement setup + Areal testing + Short measuring time + Viewing angle can vary + Defect selective + Inspection depth adjustable via the modulation frequency
Disadvantages	- Coupling medium required - Large amounts of data - Local method (scanning of the sample, which requires a lot of time) - Adjustment of the sound beam direction to the component surface - Accessibility for test head required - Complex development of test systems (probe, traversing unit, evaluation algorithm) for individual test tasks	- No depth information - Resolution decreases with increasing depth - Applicability for FRPs limited to thin structures (< 8 mm) - Homogeneous load required - Surface must reflect the light diffusely (optically rough)	- Resolution decreases with increasing depth - Applicability for FRP limited to thin structures (< 8 mm) - Homogeneous load required - Surface must absorb and emit light, preferably black surface

After completion of the tasks described in this section, milestone 3 (see Fig. 2.1) is reached and the design and manufacturing plans are available. These should be presented and critically discussed in a concluding, complete team meeting. With this, all documents necessary for the evaluation and decision in the final phase of the IPD are available.

5.5   Questions for Self-check

Below are some questions and tasks to help you reflect on the main contents of this section. The solutions can be found in Chap. 7.

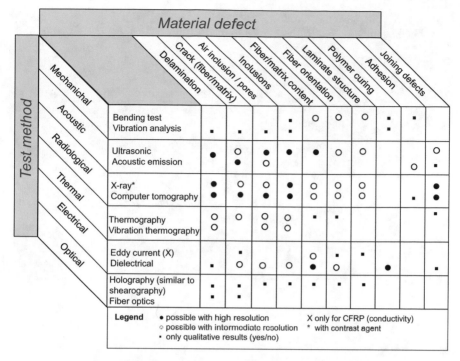

Fig. 5.36 Comparison of methods for nondestructive testing of FRP parts. Adapted from [53] (Printed with permission of Carl Hanser Verlag GmbH & Co. KG). Adapted from [117]

R33   An increasing crystallinity leads to a decreasing density and strength.— True or false?

R34   Rheological effects lead to an increase in viscosity through cross-linking.—True or false?

R35   Differential scanning calorimetry allows the determination of the glass transition temperature of a polymer.—True or false?

R36   The viscosity of thermoplastics is independent of the shear rate.—True or false?

R37   Name four criteria that should be considered when selecting a matrix polymer.

R38   Name three challenges that arise when joining FRP components.

R39   Name three methods of damage analysis for FRP components.

R40   Why does recycling of continuously fiber-reinforced polymers often lead to "downcycling"?

R41   Name four parameters that must be taken into account when procuring a production system.

R42   Name three heat transfer mechanisms and give examples from the field of FRP production.

R43   In table A, assign the semi-finished products to the FRP production process from list B.

Semi-finished product	Manufacturing process
1. Thermoset prepreg	a Resin transfer molding
2. Near-net-shape textile preform	b Compression molding ("hot pressing")
3. Sheet molding compound	c Tape laying
4. Long fiber-reinforced thermoplastic (LFT)	d Compression molding ("cold pressing")
5. Organo sheet	e Injection molding
6. Unidirectional fiber-reinforced thermoplastic tapes	f Autoclave technology
7. Unreinforced thermoplastic	g Thermoforming

Literature

1. Rimmel, O., Becker, D., Mitschang, P.: Maximizing the out-of-plane-permeability of preforms manufactured by dry fiber placement. Adv Manuf: Polym Compos Sci 2(3–4), 93–102 (2016)
2. Mack, J.: Entwicklung eines adaptiven Bebinderungsprozesses für die Preformherstellung. IVW Publication series, vol. 115. Institut für Verbundwerkstoffe, TU Kaiserslautern (2015)
3. Rao, Y., Farris, R.J.: A modeling and experimental study of the influence of twist on the mechanical properties of high-performance fiber yarns. J. Appl. Polym. Sci. 77(9), 1938–1949 (2000)
4. Goergen, C., Baz, S., Mitschang, P., Gesser, G.T.: Highly drapable organic sheets made of recycled carbon staple fiber yarns. In: 21st International Conference on Composite Materials, Xi'an, China, 20–25 Aug 2017 (2017)
5. Mitschang, P., Ogale, A., Schlimbach, J., Weyrauch, F., Weimer, C.: Preform technology: a necessary requirement for quality controlled LCM-processes. Polym. Polym. Compos. 11(8), 605–622 (2003)
6. Bitterlich, M., Ehleben, M., Wollny, A., Desbois, P., Renkl, J., Schmidhuber, S.: Tailored to reactive polyamide 6. Kunststoffe International 3(2014), 47–51 (2014)
7. Darcy, H.: Les Fontaines Publiques de la Ville de Dijon. Libraire des Corps Imperiaux des Ponts et Chausses et des Mines, Paris (1856)
8. Bruschke, M., Advani, S.G.: A finite element/control volume approach to mold filling in anisotropic porous media. Polym. Compos. 11(6), 398–405 (1990)
9. Arbter, R., Beraud, J., Binetruy, C., Bizet, L., Bréard, J., Comas-Cardona, S., Demaria, C., Endruweit, A., Ermanni, P., Gommer, F.: Experimental determination of the permeability of textiles: a benchmark exercise. Compos. A Appl. Sci. Manuf. 42, 1157–1168 (2011)
10. Vernet, N., Ruiz, E., Advani, S., Alms, J., Aubert, M., Barburski, M., Barari, B., Beraud, J., Berg, D., Correia, N.: Experimental determination of the permeability of engineering textiles: benchmark II. Compos. A Appl. Sci. Manuf. 61, 172–184 (2014)

11. Alms, J., Correia, N., Advani, S., Ruiz, E.: Experimental procedures to run longitudinal injections to measure unsaturated permeability of LCM reinforcements (2010)
12. Lundström, T.S., Stenberg, R., Bergstrom, R., Partanen, H., Birkeland, P.A.: In-plane permeability measurements: a nordic round-robin study. Compos. A Appl. Sci. Manuf. **31** (1), 29–43 (2000)
13. Grössing, H., Becker, D., Schledjewski, R., Mitschang, P., Kaufmann, S.: An evaluation of the reproducibility of capacitve sensor based in-plane permeability measurements: a benchmarking study. Express Polym Lett **9**(2), 129–142 (2015)
14. Becker, D., Grössing, H., Konstantopoulos, S., Fauster, E., Mitschang, P., Schledjewski, R.: An evaluation of the reproducibility of ultrasonic sensor-based out-of-plane permeability measurements: a benchmarking study. Adv. Manuf.: Polym. Compos. Sci. **2**(1), 34–45 (2016)
15. Berg, D., Fauster, E., Abliz, D., Grössing, H., Meiners, D., Schledjewski, R., Ziegmann, G.: Influence of test rig configuration and evaluation algorithms on optical radial-flow permeability measurement: A benchmark exercise. In: 20th International Conference on Composite Materials, Copenhagen, Denmark, 19–24 July 2015 (2015)
16. May, D., Aktas, A., Advani, S. G., Berg, D. C.: In-plane permeability characterization of engineering textiles based on radial flow experiments: a benchmark exercise. Composites Part A (submitted 24 Oct 2018), vol. 2018
17. May, D., Aktas, A., Yong, A.: International benchmark exercises on textile permeability and compressibility characterization. In: 18th European Conference on Composite Materials, Athens, Greece (2018)
18. Govignon, Q., Bickerton, S., Kelly, P.: Simulation of the reinforcement compaction and resin flow during the complete resin infusion process. Compos. A Appl. Sci. Manuf. **41**(1), 45–57 (2010)
19. Arnold, M., Broser, J., Becker, D., Mitschang, P.: Einfluss textiler Herstellungsparameter auf den maximalen Scherwinkel von Glasfasergeweben. Technische Textilien **3**(2013), 94–101 (2013)
20. Sharma, S., Sutcliffe, M., Chang, S.: Characterisation of material properties for draping of dry woven composite material. Compos. A Appl. Sci. Manuf. **34**(12), 1167–1175 (2003)
21. Christ, M., Miene, A., Moerschel, U.: Characterization of the drapability of reinforcement fabrics by means of an automated tester. In: SPE Automotive Composites Conference & Exhibition (ACCE), Troy, USA, 11–13 Sep 2012 (2012)
22. Matsudaira, M., Qin, H.: Features and mechanical parameters of a fabric's compressional property. J. Text. Inst. **86**(3), 476–486 (1995)
23. Matsudaira, M., Hong, Q.: Features and mechanical properties of fabric compressional curves. Int. J. Clothing Sci. Technol. **6**(2/3), 37–43 (1994)
24. Phoenix, S., Skelton, J.: Transverse compressive moduli and yield behavior of some orthotropic, high-modulus filaments. Text. Res. J. **44**(12), 934–940 (1974)
25. Robitaille, F., Gauvin, R.: Compaction of textile reinforcements for composites manufacturing. III: reorganization of the fiber network. Polym. Compos. **20**(1), 48–61 (1999)
26. Kelly, P., Umer, R., Bickerton, S.: Viscoelastic response of dry and wet fibrous materials during infusion processes. Compos. A Appl. Sci. Manuf. **37**(6), 868–873 (2006)
27. Chen, B., Lang, E.J., Chou, T.W.: Experimental and theoretical studies of fabric compaction behavior in resin transfer molding. Mater. Sci. Eng., A **317**(1–2), 188–196 (2001)
28. Grieser, T.: Textiles Formgebungsverhalten beim kontinuierlichen Preforming. IVW Publication series, vol. 121. Institut für Verbundwerkstoffe GmbH, TU Kaiserslautern (2016)
29. Schommer, D., Duhovic, M., Andrae, H., Steiner, K., Schneider, M., Hausmann, J.: Material characterization and compression molding simulation of CF-SMC materials in a press rheometry test. In: DGM-22, Symposium Verbundwerkstoffe und Werkstoffverbunde, Kaiserslautern, 26–28 June 2019 (2019)
30. Schürmann, H.: Konstruieren mit Faser-Kunststoff-Verbunden. Springer, Berlin (2007)

31. Mitschang, P., Neitzel, M.: Handbuch Verbundwerkstoffe. Carl Hanser GmbH & Co. KG, Munich (2004)
32. Kern GmbH: KERN RIWETA Material Selector. Downloaded from: https://www.kern.de/de/riweta, downloaded on 04 Oct 2018 (2018)
33. HUG® Technik und Sicherheit GmbH (Ergolding): Industrietechnik-Katalog der Firma HUG-Industrietechnik. Downloaded from: http://www.hug-technik.com/katalog/katalog_th.html, downloaded on 04 Oct 2018 (2018)
34. Ehrenstein, G.W., Riedel, G., Trawiel, P.: Praxis der thermischen Analyse von Kunststoffen. Hanser (2003)
35. Brown, R.: Handbook of Polymer Testing: Physical Methods. CRC Press (1999)
36. Ehrenstein, G.W.: Polymer Werkstoffe, Struktur–Eigenschaften–Anwendungen, 2nd edn. Hanser Verlag, Munich/Wien (1999)
37. Strobl, G.R.: The Physics of Polymers: Concepts for Understanding Their Structures and Behavior. Springer, Berlin/Heidelberg (1997)
38. Ehrenstein, G.W.: Faserverbund-Kunststoffe: Werkstoffe, Verarbeitung, Eigenschaften. Hanser Verlag, Munich (2006)
39. Michaeli, W., Huybrechts, D., Wegener, M.: Dimensionieren mit Faserverbundkunststoffen: Einführung und praktische Hilfen. Hanser Verlag, Munich (1995)
40. Tsai, S.W.: Introduction to Composite Materials. CRC Press, Boca Raton (1980)
41. Gay, D.: Composite Materials: Design and Applications. CRC Press, Boca Raton (2014)
42. Barbero, E.J.: Introduction to Composite Materials Design. CRC Press, Boca Raton (2017)
43. Verein Deutscher Ingenieure e.V.: VDI-Richtlinie 2014: Entwicklung von Bauteilen aus Faser-Kunststoff-Verbund (Blatt 1: Grundlagen, 1998, Blatt 2: Konzeption und Gestaltung, 1993, Blatt 3 Berechnungen, 2006) (2006)
44. European Space Agency: Structural Materials Handbook—Volume 1 Polymer Composites, ESA PSS-03-203 Issue 1, Paris (1994)
45. Arnold, M.: Einfluss verschiedener Angussszenarien auf den Harzinjektionsprozess und dessen simulative Figure. IVW Publication series, vol. 110, TU Kaiserslautern (2014)
46. Becker, D.: Transversales Imprägnierverhalten textiler Verstärkungsstrukturen für Faser-Kunststoff-Verbunde. IVW Publication series, vol. 117. Institut für Verbundwerkstoffe GmbH, Technische Universität Kaiserslautern (2015)
47. Menges, G., Geisbüsch, P.: Die Glasfaserorientierung und ihr Einfluß auf die mechanischen Eigenschaften thermoplastischer Spritzgießteile—Eine Abschätzmethode. Colloid Polym. Sci. **260**(1), 73–81 (1982)
48. e-mobil BW, Landesagentur für Elektromobilität und Brennstoffzellentechnologie; Fraunhofer IPA, Ministerium für Finanzen und Wirtschaft Baden-Württemberg, Ministerium für Wissenschaft, Forschung und Kunst Baden-Württemberg *Spanende Bearbeitung von Leichtbauwerkstoffen,,* Downloaded from: http://publica.fraunhofer.de/dokumente/N-226391.html, downloaded on 11 Sep 2018 (2012)
49. Noll, T.J.: Beitrag zur Entwicklung punktueller Lasteinleitungen und Verbesserung der Versagensanalyse für Faser- Kunststoff-Verbund-Strukturen unter zyklischer Belastung. IVW Publication series, vol. 81. Institut für Verbundwerkstoffe GmbH, TU Kaiserslautern (2008)
50. Albert, C., Fernlund, G.: Spring-in and warpage of angled composite laminates. Compos. Sci. Technol. **62**(14), 1895–1912 (2002)
51. Kappel, E.: Process Distortions in Composite Manufacturing—From an Experimental Characterization to a Prediction Approach for the Global Scale. Otto-von-Guericke-Universität, Magdeburg, Thesis (2013)
52. Kleineberg, M.: Präzisionsfertigung komplexer CFK-Profile am Beispiel Rumpfspant, DLR Forschungsbericht: ISSN 1434-8454. Technische Universität Carolo-Wilhelmina zu Braunschweig, Thesis (2009)
53. Neitzel, M., Mitschang, P., Breuer, U.: Handbuch Verbundwerkstoffe: Werkstoffe, Verarbeitung, Anwendung. Carl Hanser Verlag GmbH Co KG, Munich (2014)

54. Ernstberger, U., Weissinger, J., Frank, J.: Mercedes-Benz SL: Entwicklung und Technik. Springer Fachmedien Wiesbaden (2013)
55. Schreckenberger, H.: Vermeidung von Kontaktkorrosion im Leichtbau bei der Anbindung von Magnesium, Aluminium und CFK. 39. Ulmer Gespräch—Forum für Oberflächentechnik der DGO (Deutsche Gesellschaft für Galvano- und Oberflächentechnik e.V.), Neu-Ulm, 17–18 May 2017 (2017)
56. Francis, R.: Galvanic Corrosion: A Practical Guide for Engineers. NACE—National Association of Corrosion Engineers (2001)
57. Schreckenberger, H.: Risiko der Kontaktkorrosion bei CFK-Bauteilen. WOMag: Kompetenz in Werkstoff und funktioneller Oberfläche, vol. 4/2013 (2013)
58. Kalpakjian, S., Schmid, S.R., Werner, E.: Werkstofftechnik. Pearson Studium, London (2011)
59. Liston, E., Martinu, L., Wertheimer, M.: Plasma surface modification of polymers for improved adhesion: a critical review. J. Adhes. Sci. Technol. 7(10), 1091–1127 (1993)
60. Brockmann, W., Geiß, P.L., Klingen, J., Schröder, K.B.: Klebtechnik: Klebstoffe, Anwendungen und Verfahren. Wiley, Weinheim (2012)
61. Molnár, P.: Stitching technique supported preform technology for manufactoring fiber reinforced polymer composites. IVW Publication series, vol. 74. Institut für Verbundwerkstoffe GmbH, TU Kaiserslautern (2007)
62. Molnar, P., Mitschang, P., Felhos, D.: Improvement in bonding of functional elements with the fiber reinforced polymer structure by means of tailoring technology. J. Compos. Mater. 41(21), 2569–2583 (2007)
63. Marcotodo Wärmetechnik: Das TeQua® Cure System - Reparatur und gezielte Aushärtung von Bauteilen aus Faserverbundwerkstoffen. Vortrag am 09.04.2013 im Gründerzentrum der Neue Materialien Bayreuth GmbH (2013)
64. Holtmannspötter, J., Feucht, F., Meyer, J.C., de Freese, J.D., von Czarnecki, J.: Schnelle und zuverlässige Reparatur. adhäsion KLEBEN & DICHTEN 57(11), 36–41 (2013)
65. Breuer, U.P.: Commercial Aircraft Composite Technology. Springer, Berlin (2016)
66. Europäische Union: Richtlinie 2008/98/EG des Europäischen Parlaments und des Rates der Europäischen Union vom 19. November 2008 über Abfälle und zur Aufhebung bestimmter Richtlinien. Amtsblatt der Europäischen Union (2008)
67. Kreibe, S.: CFK-Recycling, Ökologie und Abfallwirtschaft: Ein Blick über den Tellerrand. Fachtagung Carbon Composites, Augsburg, 1–2 Dec 2015 (2015)
68. Kühnel, M.: The global CFRP market 2016. In: Conference: Experience Composites, Augsburg, 21 Nov 2016 (2016)
69. Witten, E., Sauer, M., Kühnel, M.: Composites-Marktbericht 2017: Marktentwicklungen, Trends, Ausblicke und Herausforderungen. In: Ed. AVK Industrievereinigung verstärkte Kunststoffe e.V./Carbon Composites e.V. (2017)
70. Bundesrepublik Deutschland: Kreislaufwirtschaftsgesetz (Gesetz zur Förderung der Kreislaufwirtschaft und Sicherung der umweltverträglichen Bewirtschaftung von Abfällen), became effective on 01.03.2012 bzw, 01 June 2012 (2012)
71. Schiebisch, J.: Zum Recycling von Faserverbundkunststoffen mit Duroplastmatrix. Universität Erlangen-Nürnberg, Thesis at Institute of Polymer Technology (1996)
72. Stelzer, G.: Zum Faser- und Eigenschaftsabbau bei Verarbeitung und Recycling diskontinuierlich faserverstärkter Kunststoffe: Anwendung des Mikrobiegeversuchs zur Faserfestigkeitsbestimmung am Beispiel methodischer Untersuchungen des Eigenschaftsabbaus. Thesis, TU Kaiserslautern (2003)
73. Carbon Composites e.V.: Sonderheft 2016: Jahresthema "Recycling" (2016)
74. Duhovic, M., Romanenko, V., Schommer, D., Hausmann, J.: Material characterization of high fiber volume content long fiber reinforced SMC materials. In: 14th International Conference on Flow Processes in Composite Materials, Lulea, Sweden, 30 May–01 June 2018 (2018)
75. Schommer, D., Duhovic, M., Hausmann, J.: Development of a Solid Mechanics Based Material Model Describing the Behavior of SMC Materials. In: 14th International

Conference on Flow Processes in Composite Materials, Lulea, Sweden, 30 May–01 June 2018 (2018)
76. Goergen, C., Baz, S., Mitschang, P., Gresser, G.T.: Recycled carbon fibers in complex structural parts-organic sheets made of rCF staple fiber yarns. Key Eng. Mater. **742**, 602–609 (2017)
77. Kreibe, S., Hartleitner, B., Gottlieb A., Berkmüller, R., Förster, A., Tronecker, D., Reinelt, B., Wambach, K., Rommel, W.: Bifa Umweltinstitut GmbH: Final report MAI Recycling (2015)
78. Fiberline Composites (press release from 15.09.2010): Durchbruch im GFK-Recycling. Downloaded from: https://fiberline.de/news/miljoe/durchbruch-im-gfk-recycling, downloaded on 20 Sep 2018
79. Teipel, U., Seiler, E.: Recycling von Kompositebauteilen aus Kunststoffen als Matrixmaterials. Projektverbund ForCycle: project flyer (2014)
80. Pahl, G., Beitz, W., Schulz, H.-J., Jarecki, U.: Pahl/Beitz Konstruktionslehre: Grundlagen erfolgreicher Produktentwicklung. Methoden und Anwendung. Springer, Berlin/Heidelberg (2013)
81. Bhattacharyya, D., Fakirov, S.: Synthetic Polymer-Polymer Composites. Carl Hanser Verlag GmbH & Company KG, Munich (2012)
82. Dzalto, J.: Entwicklung eines großserientauglichen Aufheizprozesses für naturfaserverstärkte Kunststoffe. IVW Publication series, vol. 117. Institut für Verbundwerkstoffe GmbH, TU Kaiserslautern (2018)
83. Von Böckh, P., Wetzel, T.: Wärmeübertragung: Grundlagen und Praxis, 2nd edn. Springer, Berlin/Heidelberg (2017)
84. Moser, L.: Experimental analysis and modeling of susceptorless induction welding of high performance thermoplastic polymer composites. IVW Publication series, vol. 101. Institut für Verbundwerkstoffe GmbH, TU Kaiserslautern (2012)
85. Wiegmann, A., Zemitis, A.: EJ-HEAT: A fast explicit jump harmonic averaging solver for the effective heat conductivity of composite materials. Reports of the Fraunhofer Instituts für Techno- und Wirtschaftsmathematik (94) (2006)
86. Mark, J.E.: Physical Properties of Polymers Handbook. Springer, New York (2007)
87. Baehr, H.D., Stephan, K.: Wärme-und Stoffübertragung. Springer, Berlin/Heidelberg (1998)
88. Langeheinecke, K., Jany, P., Thieleke, G.: Thermodynamik für Ingenieure: ein Lehr- und Arbeitsbuch für das Studium. Vieweg + Teubner Verlag, Wiesbaden (2011)
89. Neumann, S.P.: Theoretical derivation of Darcy's law. Acta Mech. **3–4**, 153–160 (1977)
90. Zierep, J., Bühler, K.: Grundzüge der Strömungslehre. Springer, Berlin/Heidelberg (2013)
91. Das, B.M.: Advanced Soil Mechanics. CRC Press, Baco Raton (2013)
92. Drapier, S., Pagot, A., Vautrin, A., Henrat, P.: Influence of the stitching density on the transverse permeability of non-crimped new concept (NC2) multiaxial reinforcements: measurements and predictions. Compos. Sci. Technol. **62**(15), 1979–1991 (2002)
93. Bear, J.: Dynamics of Fluids in Porous Media. American Elsevier, New York (1972)
94. Shojaei, A., Trochu, F., Ghaffarian, S., Karimian, S., Lessard, L.: An experimental study of saturated and unsaturated permeabilities in resin transfer molding based on unidirectional flow measurements. J. Reinf. Plast. Compos. **23**(14), 1515–1536 (2004)
95. Michaeli, W., Hammes, V., Kirberg, K., Kotte, R., Osswald, T.A., Specker, O.: Process Simulation in the RTM Technique. Kunststoffe-German Plastics **79**(8), 739–742 (1989)
96. Trochu, F., Gauvin, R., Gao, D.M.: Numerical analysis of the resin transfer molding process by the finite element method. Adv. Polym. Technol. **12**(4), 329–342 (1993)
97. Michaud, V., Grajzgrund, H.J., Manson, J.A.E.: Influence of preform compressive behavior in liquid composite molding. In: 5th International Conference on Composite Materials, San Diego, USA, 29 July–1 Aug 1999 (1999)
98. Mayer, C.: Prozessanalyse und Modellbildung bei der Herstellung gewebeverstärkter, thermoplastischer Halbzeuge. IVW Publication series, vol. 5. Kaiserslautern, IVW GmbH (2000)

99. Ahn, K.J., Seferis, J.C., Berg, J.C.: Simultaneous measurements of permeability and capillary-pressure of thermosetting matrices in woven fabric reinforcements. Polym. Compos. **12**(3), 146–152 (1991)

100. Batch, G.L., Chen, Y.-T., Macoskot, C.W.: Capillary impregnation of aligned fibrous beds: experiments and model. J. Reinf. Plast. Compos. **15**(10), 1027–1051 (1996)

101. Hopmann, C., Michaeli, W., Greif, H., Wolters, L.: Technologie der Kunststoffe. Carl Hanser Verlag GmbH & Co. KG, Munich (2015)

102. Christmann, M.: *optimierung der organoblechherstellung durch 2d-imprägnierung*. IVW Publication series, vol. 114. Institut für Verbundwerkstoffe GmbH, TU Kaiserslautern (2014)

103. Trochu, F., Gauvin, R.: RTMFLOT—an integrated software environment for the computer simulation of the resin transfer molding process. J. Reinf. Plast. Compos. **13**(3), 262–270 (1994)

104. Walbran, W.A., Verleye, B., Bickerton, S., Kelly, P.A.: RTM and CRTM simulation for complex parts. In: Processing and Fabrication of Advanced Materials XIX, Auckland, New Zealand, 14 Feb 2011 (2011)

105. Koorevaar, A.: Simulation of liquid injection molding. In: 23rd SAMPE Europe Conference, Paris, 09–11 Apr 2002 (2002)

106. Simacek, P., Advani, S.G., Binetruy, C.: LIMS: a comprehensive tool to design, optimize and control the filling process in liquid composite molding. JEC—Compos. **8**, 143–146 (2004)

107. Beresheim, G.: Thermoplast-Tapelegen: Ganzheitliche Prozessanalyse und -entwicklung. IVW Publication series, vol. 32. Institut für Verbundwerkstoffe GmbH, TU Kaiserslautern (2002)

108. Advani, S.G., Sozer, E.M.: Process Modeling in Composites Manufacturing, 2nd edn. CRC Press, Boca Raton (2010)

109. Tietjen, T., Müller, D.H.: FMEA-Praxis: das Komplettpaket für Training und Anwendung. Hanser Verlag, Munich (2003)

110. Gryzagoridis, J., Findeis, D.: Tap testing of composites benchmarked with digital shearography. Insight-Non-Destructive Test. Condition Monit. **56**(1), 35–38 (2014)

111. Cawley, P., Adams, R.: The mechanics of the coin-tap method of non-destructive testing. J. Sound Vib. **122**(2), 299–316 (1988)

112. Summerscales, J.: Non-Destructive Testing of Fiber Reinforced Plastics Composites. Springer, Netherlands (1990)

113. Schuth, M., Buerakov, W.: Handbuch Optische Messtechnik: Praktische Anwendungen für Entwicklung, Versuch, Fertigung und Qualitätssicherung. Carl Hanser Verlag GmbH & Co. KG, Munich (2017)

114. Fuhrmann, E.: Einführung in die Werkstoffkunde und Werkstoffprüfung: Werkstofe: Aufbau - Behandlung - Eigenschaften. Expert-Verlag, Renningen (2008)

115. Kelkel, B., Popow, V., Gurka, M.: Combining acoustic emission with passive thermography to characterize damage progression in cross-ply CFRP laminates during quasi-static tensile loading. In: 12th European Conference on Non-Destructive Testing, Gothenburg, Sweden, 11–15 June 2018

116. Popow, V., Budesheim, L., Gurka, M.: Comparison and evaluation of different processing algorithms for the nondestructive testing of fiber reinforced plastics with pulse thermography. Mater. Test. **60**(6), 607–613 (2018)

117. von Wachter, F.K.: Ein Beitrag zur Automatisierung der zerstörungsfreien Ultraschallprüfung von Faserverbundkunststoffen. Shaker, Aachen (1992)

Chapter 6
Phase 4: Evaluation and Decision

Abstract This section deals with the last step of the product development, which includes the selection of one of the elaborated designs for actual realization. To form a basis for this selection, all alternatives must be evaluated, whereby a holistic evaluation requires economic, technical and strategic aspects to be taken into account. This section therefore contains

- methods for economic evaluation,
- methods for comparison of the achieved component properties with the requirements catalog,
- fundamentals of prototype design and testing, and finally
- a procedure for overall techno-economic and strategic assessment of the alternatives.

6.1 Overview

Figure 6.1 shows an overview of the procedure followed in phase 4. It comprises evaluation of all alternatives and the decision about which one should be finally implemented.

6.2 Economic Evaluation

As with all other material groups, FRP components must usually meet specific economic requirements in order to be considered for an industrial implementation. The demand for "economic efficiency" is usually at the top of the requirements list.

Economic efficiency is a formal principle that forms the basis of any economic activity. It can also be defined as a key figure, calculated as the quotient of income and expenditure or output and costs (cost efficiency). If the ratio is equal to one, the costs are covered. If it is greater, a profit is made. If cost efficiency is achieved in the production of a FRP component, this means that the production costs are less than or equal to the added value of the production [1, 2].

© The Author(s), under exclusive license to Springer Nature Switzerland AG 2021   217
D. May, *Integrated Product Development with Fiber-Reinforced Polymers*,
Engineering Materials, https://doi.org/10.1007/978-3-030-73407-7_6

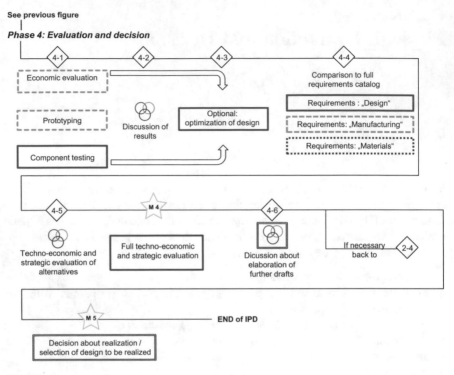

Fig. 6.1 Overview of phase 4—evaluation and decision

In addition to an absolute economic efficiency, a relative economic efficiency can be defined as cost-related superiority of a process over other, competing processes. In mathematical terms, a process A is relatively economic efficient compared to a process B, if the following applies, while the added value is equal [3]:

$$\text{Costs}(A) \div \text{Costs}(B) < 1 \qquad (6.1)$$

In this case, comparing the economic efficiency equals comparing the costs. However, there may also be changes concerning the revenues, e.g. an added value due to weight savings. Hence, different substitution strategies can be relevant. Defining the substitution strategy, at an early stage of product development, is particularly important when FRPs compete with other materials. This is the only way to ensure that the entire IPD is aligned with this strategy. Figure 6.2 shows different substitution strategies on the example of a decision between a FRP- and a metal-based solution, respectively. Nevertheless, all following statements apply for decisions between any two technical alternatives.

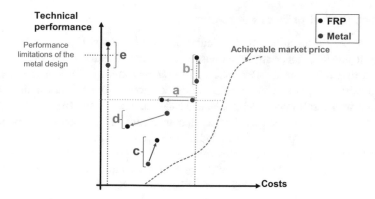

Fig. 6.2 Substitution strategies for FRP parts

The decision will be in favor of the FRP solution, if one of the following applies:

(a) For the **same performance** level, the FRP variant is **cheaper** than the metal variant.
(b) At the **same costs**, the FRP variant offers a **higher performance** level, (illustrated by the greater distance to the curve marking the achievable market price).
(c) At **higher costs**, the FRP variant offers a **higher performance** level, which leads to higher revenues on the market, whereas the higher revenues more than compensate for the higher costs (increase in economic efficiency).
(d) The FRP variant offers a **lower performance** level, combined with **lower costs**, which more than compensate for the loss of revenue.
(e) The **required performance** level **cannot be achieved** with the metal version. In addition to the common target of lightweight design, also requirements such as corrosion resistance, tribological resistance, transparency for electromagnetic radiation, design requirements, dimensional accuracy, or annual production quantities may also be decisive, especially with regard to FRPs.
(f) Even if none of the cases shown in the diagram applies, **strategic or political decisions** can lead to the selection of the FRP variant. E.g. when there is an intention to generate expertise, or to improve the image, or when there is an expectation for decreasing costs in the future.

The cases (c) and (e) are typical for high performance FRPs (continuous fiber reinforcement). Additionally, cases (a), (b) and (d) are quite common for short and long fiber-reinforced polymers.

For an economic assessment, the following fundamental questions must now be answered:

- How high are the **manufacturing costs** (depending on the planned annual output at product launch and later in the future)?
- How high are the **investment** costs?

- Which **market price** can be achieved?

Of course, the accuracy of a calculation-based answer to the last question is limited. It must be rather answered based on market research or experience. If it is assumed that the same market price can be achieved with all considered alternatives, a cost comparison is sufficient. One approach to cost calculation will be presented in the following section, **process-based cost modeling**.

There are numerous approaches for the modeling of product costs, which e.g. differ concerning the phases of the product life cycle that are considered or the level of detail. At this point, a very effective method is presented. It gives meaningful results even in an early development phase, and it does not require extensive training or knowledge to be applied. This method is process-based cost modeling (PBCM). Here, the elements that influence costs are individually derived from engineering principles and the technical parameters of the manufacturing process. The corresponding basic procedure is shown in Fig. 6.3, which was adapted from [4].

The basis for the modeling is, among other things, a derivation of the process cycle time, which is decisive for the costs per unit. The principle applicability of PBCM to FRP production processes was demonstrated by Jens Schlimbach [4], whose work forms the basis for the methodology presented here. The PBCM method is based on the principles of investment and cost accounting. A statement about the costs per unit is made by allocating all costs, including fixed costs, to the output quantity. Accordingly, the total costs (C_{tot}) for the annual production (p) of a component are the sum of all fixed and variable costs (C_f and C_v, respectively), resulting from the production:

$$C_{\text{tot}} = C_f + C_v \qquad (6.2)$$

The costs per unit (c) are then calculated based on the annual production, whereas a simple calculation can be achieved, when full utilization of the used production systems is assumed. This way, the fixed costs can be allocated to the annual production capacity:

$$c = C_{\text{tot}}/p \qquad (6.3)$$

The **fixed costs** are calculated taking into account cycle time, annual working days and shift planning. The fixed costs for a production unit j result from the respective interest costs for bound capital (IC_j), the annual depreciation (D_j), the space costs (SC_j) and the maintenance costs (MC_j):

$$C_f = \sum_j (D_j + IC_j + SC_j + MC_j) \qquad (6.4)$$

The individual fixed cost components can be calculated taking into account the investment costs Inv_j, the number of periods of use n_j, the residual value at the end of the use life RV_{jn}, the residual value before the last year of use RV_{jn-1} and the

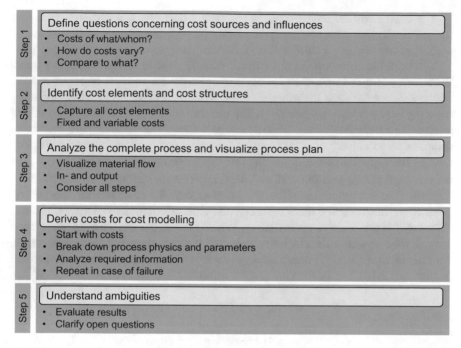

Fig. 6.3 Procedure of process-based cost modeling. Adapted from [4]

required area A_j. To do this, meaningful assumptions concerning economic boundary conditions must be made. This concerns the interest rate i as well as the room and maintenance cost rate, cr_S and cr_M, respectively:

$$D_j = \frac{In_j - RV_{jn}}{n_j} \tag{6.5}$$

$$IC_j = i \cdot \frac{In_j + RV_{jn-1}}{2} \tag{6.6}$$

$$SC_j = cr_S \cdot A_j \tag{6.7}$$

$$MC_j = cr_M \cdot \frac{In_j + RV_{jn-1}}{2} \tag{6.8}$$

For further simplification, no interperiodic monetary value effects[1] are taken into account, and average values are used for the cost of capital.

---

[1]For every investment, there are opportunity costs, because the money could have been invested in other alternatives or because additional costs such as interests emerge. Therefore, the costs incurred over the lifetime of a system are not to be assessed equally and cannot simply be added

The **fixed costs** are independent of the output quantity. If a system is procured for the production of a certain type of component, the costs will be the same, regardless whether ten or one hundred thousand components are produced per year.

The **variable costs**, on the other hand, directly depend on the number of units. With each additional unit, the total costs increase by the variable costs per unit c_v, which includes the costs for materials (c_M), labor (c_L), energy (c_E), other operating costs (c_O) and production waste (c_W). The material costs can be calculated by taking into account the quantity m_{Cl} of each material l required for each part, the corresponding material price cr_{Ml} for each used material l, as well as the production loss rate q_d, the rejection rate q_{Rej} and the recycling rate q_{Rec}. For the labor costs, the number of staff A_{NW} and the staff cost rate cr_{Ls} for each staff member s involved is required. The energy consumption W_A together with the cost rates for energy cr_E, production waste cr_W and other costs cr_O allow the calculation of energy, waste and other costs, if the effective cycle time t_{eff} is also known. For the latter, [4] developed a cycle time model with which the times of highly automated processes can be determined heuristically. Further approaches to cycle time estimation are described in Sect. 5.5.2.1.

$$c_v = c_M + c_L + c_E + c_O + c_W \tag{6.9}$$

$$c_M = \sum_l m_{Cl} \cdot \left(1 + q_d + q_{Rej} - q_{Rec}\right) \cdot cr_{Ml} \tag{6.10}$$

$$c_L = \sum_s t_{eff} \cdot A_{NW} \cdot cr_{Ls} \tag{6.11}$$

$$c_E = t_{eff} \cdot W_A \cdot cr_E \tag{6.12}$$

$$c_O = t_{eff} \cdot cr_O \tag{6.13}$$

$$c_W = \sum_l m_{Cl} \cdot \left(q_d + q_{Rej} - q_{Rec}\right) \cdot cr_W \tag{6.14}$$

In accordance with the previously described equations, Fig. 6.4 shows the development of total costs and costs per unit as a function of the annual output quantity. The total costs initially increase linearly with increasing output. Thereby the gradient results from the variable costs that have to be spent for each additional unit. The curve begins at the level of the fixed costs, which remain constant in this first phase. This results in a degressive trend for the costs per unit (unit cost degression), as the fixed costs are distributed over an ever increasing number of units. A change only occurs when a capacity limit is reached, i.e. an increase in the annual output quantity is no longer possible with the given production systems, as

---

together. Simply put, costs must be evaluated higher, the earlier they appear, because the money is bound earlier. Such effects are neglected in the method presented here [4, 5].

these are working at full capacity. Then a further set of production systems must be purchased, resulting in a jump in fixed costs, which is reflected in a corresponding jump in costs per unit. If it is not possible to utilize the production systems to full capacity, the costs per unit are higher than before, but the degressive course eventually leads to costs equal to or even lower than before the jump [4].

It is important to keep these relationships in mind, because often it can be derived directly, without further consideration, whether a process can be economically meaningful or not. For example, with an annual production quantity of one thousand units, a process requiring a press with an investment of 1 million € will not make sense (unless other products can be manufactured with the press), because the annual fixed costs would only be distributed among the one thousand units, which means that the costs per unit are enormous. It is also important to estimate future developments: What annual output quantity is planned one, two or five years after the start of production? This question has an enormous influence on whether procurement of a production system is economically meaningful.

If shift work is used, further changes may occur. For example, it is possible that the annual output quantity is first increased by switching from one-shift to two- or three-shift operation, when a capacity limit is reached. This, in turn, may result in higher labor costs (shift surcharge), which will increase variable costs.

Now, in order to calculate the costs related to a specific manufacturing process and process chain, respectively, first of all, it is necessary to identify the individual elements the costs can be allocated to (see Fig. 6.3, steps 2 to 4). For this purpose, the targeted manufacturing process is broken down step by step, as shown in Fig. 6.5. Each process is divided in process steps, which in turn can be divided into activities. Finally, the elements involved can be assigned to these activities. Such elements are the production systems (see Sect. 5.5.1), the tools and the semi-finished products or input materials. According to the equations already explained, costs can be assigned to the elements. It is important to make sure that there are no redundancies. For example, in a thermoforming process the organo sheet will appear as an element when considering many activities. However, since it is always the same organo sheet, its material costs of course only are to be included once in the unit price.

The total costs C_i for a process i involved in the product creation can thus be calculated from m_P (the frequency of process step P), m_k (the frequency of activity k), m_l (the frequency of element l) and V_{Pkl} (the single value of element l of activity k of process step P) [4, 7]:

$$C_i = \sum_P \sum_k \sum_l (m_P \cdot m_k \cdot m_l \cdot V_{Pkl}) \tag{6.15}$$

Summing up all the costs for all processes involved in product creation results in the manufacturing costs for a single unit with a specific manufacturing concept. This way, all designs can be compared in terms of their process and cost structure. For this purpose, it is advisable to implement the equations shown in the previous section in a spreadsheet. This allows efficient execution of a sensitivity analyses, for

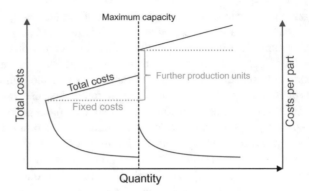

Fig. 6.4 Cost development depending on production quantity. Adapted from [4] (Image adapted, printed with permission of Leibniz-Institut für Verbundwerkstoffe GmbH)

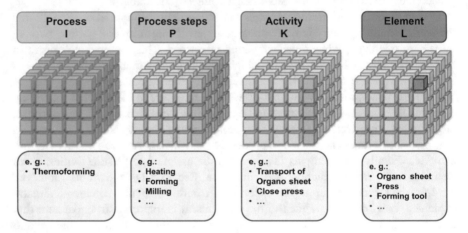

Fig. 6.5 Process hierarchy and data structure of the model [6]. Image adapted, printed with permission of Leibniz-Institut für Verbundwerkstoffe GmbH

example, with regard to the relationship between the number of units, the unit price and the effects of an increase in the energy cost rate. If additionally, the process-related equations, especially those forming the basis for cycle time estimation, are implemented, it is relatively easy to identify starting points for process and design optimization. By changing just a few values, it is then e.g. possible to make a statement, whether the purchase of a more expensive but also more powerful production system (e.g. faster tape layer or larger oven) is worthwhile. Another important point concerns the question of the economically optimal number of the respective production systems. Since individual process steps can be parallelized in the context of series production, it can make sense to use multiple systems of the same type in order to prevent bottlenecks and increase capacity utilization. An example has already been shown in Sect. 5.5.2. During the sensitivity analysis, it can now be determined, which system configuration is actually the most economical one.

With regard to the initially defined questions, it is important to be clear about the objective of the cost modeling from the beginning. For highly accurate cost predictions, accurate input data regarding investments and cycle times is required. This in turn requires quite an effort, because then e.g. process simulations are applied for cycle time prediction instead of rough estimations. This effort is only justified if such an accurate cost prediction is actually needed. In many cases, at this stage of development, the objective will rather be to compare the economic potential of different component and process designs. In such a comparative consideration, several designs are based on the same assumptions. For a meaningful relative comparison, it is therefore often sufficient if these assumptions are based on rather roughly estimated data. Furthermore, identification of the basis cost structure, as a basis for a sensitivity analysis with the target to identify optimization potential, does also not require data accurate to the cent. Hence, the objective defines the requirements for the input data and should be clearly defined.

6.3   Prototyping and Component Testing

Nowadays, numerical simulations can already predict the final component properties quite well. Nevertheless, due to the multitude of influences on the actual properties, induced by materials, semi-finished products and processes, production and testing of prototypes is still indispensable. This is especially true, when it comes to making a final decision about the realization of a product. With the production and testing of prototypes, several objectives can be simultaneously pursued:

1. Proof of the **technical performance** of components (especially proof of the load-bearing capacity of structural components).
2. Validation of **material suitability** (e.g. weather resistance).
3. Verification of the **suitability of semi-finished products** (e.g. drapeability/impregnability of textiles).
4. Checking the achievable **quality** (e.g. dimensional accuracy, porosity in the laminate, etc.).
5. Evaluation of the **suitability of the selected process** and the achievable cycle times.

Especially for the last two points it is important that the process used for prototype manufacturing comes as close as possible to the final intended manufacturing process. Using exactly the same process is often not possible due to economic reasons. Particularly processes that are more suitable for larger quantities would cause high costs, even if the equipment is not procured directly, but the production is ordered externally. This is mainly due to the often-expensive tools. It is therefore common to initially use alternative processes, which allow the target materials (not necessarily the target semi-finished products) to be processed in a small-batch production. It must be taken into account that in this case there will be deviations from the target process, especially with regard to the final component properties and quality. A typical example is prototype production by vacuum infusion (VI), if an RTM process is planned as the final process. Since no closed tool is used in VI compared to RTM, differences can occur with regard to the fiber volume content, even if the same or similar materials are used. First prototypes are often produced by autoclave technology, as it is very flexible regarding the materials and the component geometry. However, it must be considered that the component quality is often significantly higher than in the actual target process. Generally, the use of substitute processes is economical and technically meaningful. However, it is important to be fully aware of the differences between substitute and target processes when evaluating the test results. Therefore, the team members have to define the objective of the prototype testing together, so that the team member from the field of manufacturing technology can select a suitable process and inform the other team members about the differences. This way, they can properly design their tests on the prototypes and interpret the results correctly.

6.4   Optional: Design Optimization

Following the economic process analysis and prototype testing, the results are discussed in a team meeting (steps 4–2 in Fig. 6.1). One of the main points of discussion should be to decide, based on the findings, whether it makes sense to revise the design (step 4–3) before the final evaluation is made. This may be useful, for example, …

- … if the cost analysis suggests that minor design changes would have a major impact on the economic efficiency. For example, because a certain standard semi-finished product can be used if the component width is only slightly reduced.
- … if the prototype test shows major deviations from the FE simulation, which induces that the design is based on wrong assumptions.

Especially when several designs are compared at this point, it should be considered whether the revision should be carried out directly. If there is no doubt about feasibility of the optimizations, the consequences for the final evaluation can be estimated if necessary. This way, it is avoided that too much effort is put into designs, which may not be further pursued in the end.

6.5   Final Comparison with the Requirements Catalog

In order to prepare the overall evaluation, all alternatives should be checked with regard to the requirements catalog. This comparison of target characteristics and reached characteristics is required so that the overall evaluation is holistic. It is advisable that at first, each team member focuses on the list of requirements that was defined for the respective area and adds statements on the achievement of targets for each requirement and each alternative. For quantified requirements (e.g. weight minimization), the target achievement corresponds to a number (e.g. the achieved weight in kg). For qualitative requirements (e.g. minimum implementation risk), it should first be described to what extent an alternative contributes to or contradicts the achievement. An evaluation and quantification will only take place in the course of the overall evaluation.

Once completed, the area-specific comparisons are then merged. While economic and technical questions can be widely answered by cost modeling and component testing, the requirements catalog may also contain strategic aspects that consider product development in a larger context (see Sect. 3.5.3). In some cases, these questions can only be answered by statistical methods that are subject to an inherent uncertainty. Corresponding methods can be found, for example, in [8] and [9] and will not be considered in detail here. Even if these sophisticated methods are not applied, due to the required effort, it is essential that all the above issues are taken into account, in order to make a holistic evaluation. Once again, it is the integration of different expertise in the development team, which ensures that profound answers to all questions can be found for all designs. Only the integration ensures that the necessary experience and expertise is available. Hence, merging and completion of the separate evaluations should take place in a topic-specific meeting, in which all team members participate (steps 4–4 in Fig. 6.1).

6.6 Overall Techno-Economic and Strategic Evaluation

The final step is the overall evaluation. This may also include product developments that were not considered in the previous section, for example, a competing product development in metal design. For the final evaluation, simple decision-making tools can be a helpful support. E.g. the benefit analysis, which is especially interesting, because it can also be used as a checklist to ensure that all aspects are taken into account. Applying the principles of IPD (especially teamwork) is of high relevance for an efficient overall evaluation, where subjective influences are prevented. In the following section, a corresponding methodology for the benefit analysis is proposed. It allows a relatively fast but holistic evaluation. More detailed explanations on the benefit analysis and alternative evaluation methods can be found e.g. in [8–10]. The single steps of the proposed method are the following (according to [9]):

1. The first step of the benefit analysis is the **definition of the target criteria**. Table 6.1 lists target criteria that are proposed as the most important ones. This list can be extended as required, taking into account the requirements catalog. For example, the technical achievement of objectives can be further divided into lightweight design objectives, processing objectives, etc. With increasing differentiation, however, it becomes increasingly important to ensure that the target

Table 6.1 Target criteria for a benefit analysis for FRP products

Manufacturing costs	Corresponds to the sum of all costs attributable to the component during production. If the same sales value (market price) is expected for all alternatives, a comparison of the production costs is sufficient to evaluate the potential profit
Investment costs	Mainly includes the required production systems
Development potential (serial size)	If the annual output is expected to change, for example, due to increasing market size, this may affect the economic efficiency of the production process. It should therefore be assessed whether the potential of the manufacturing process is in line with the expected development of the annual output
Achievement of technical targets	Corresponds to the technical assessment (see Sect. 6.6). Provided the list of requirements has been carefully prepared, alternatives that do not meet the minimum requirements should not be considered. At this point, therefore the benefits of the alternatives that go beyond these minimum requirements (e.g. the fulfillment of wishes or a weight reduction beyond the minimum target) are assessed
Implementation risk	Defines the height of the risk that the targeted technical performance cannot be achieved
Development effort and duration	Corresponds to an assessment of the time required for implementation (often critical, as a fast market introduction is to be achieved) and development costs as well as the necessary personnel deployment, etc.
Strategic benefit	Summarizes the benefits of an alternative that exceed the expected profits associated with the product. For example, advertising effects that affect the sales of other products of the manufacturer, or possibilities to transfer the developed technologies to other products. A good example for this is small-series sports cars in the fleets of car manufacturers, who otherwise tend to sell cars for the mass market. These sports cars serve as both technology carriers and advertising measures
Characteristics of use	Summarizes all aspects relevant in the use phase. This especially includes the question to which extent the product is designed for maintenance and repair
Sustainability	Assessment of the environmental impact of the product including recyclability

criteria are independent of each other so that there is no unintended multiple weighting. For example, it would be wrong to use the profit per component and the manufacturing costs per component as target criteria, since the costs are already included in the profit.

2. In a second step, the **target criteria are weighted**. Here too, various methods can be found in the relevant literature. A quite effective method is the preference matrix method. Here, all criteria are compared pairwise with regard to their relevance. Then, for each criterion, it is counted how often it was considered more important compared to another criteria. Using the frequencies, a ranking list is created, in which the criterion with the highest number of "wins" has the

highest rank. If two or more criteria have the same frequency, they all receive the same rank. The criterion with the next lower frequency then still gets the rank that would have resulted if ranking had been numbered consecutively (i.e. rank numbers are skipped). Then the reverse order of priority is created. For each criterion, the weight is calculated, by dividing the sum of the weights by the sum of the ranks and then multiplying this value with the respective inverse rank. It simplifies the procedure if the sum of the weights is set to 100. Table 6.2 shows the preference matrix and an example weighting for the proposed target system. Here, the letter of the respective "winner" criterion of a comparison is written in the preference matrix. The individual fields always refer to the comparison between the criteria along the diagonal. A somewhat more sophisticated method is offered by the analytical hierarchy process (AHP). In AHP it is not only determined which criterion is more important, but also how much more important [11].

Table 6.2 Target system and preference matrix for target weighting

	Target criteria	Preference matrix					Frequency in preference matrix	Rank	Inversed rank	Weighting
a	Market price—manufacturing costs						8	1	9	21.43
		a								
b	Investment costs		a				6	2	6	14.29
			b	a						
c	Development potential		b	a			4	5	5	11.90
			c	b	a					
d	Technical target achievement		e	f	a		3	6	4	9.52
			e	f	b	a				
e	Implementation risk		f	c	b	a	6	2	6	14.29
			e	d	c	b				
f	Development effort and duration		e	d	c		6	2	6	14.29
			f	e	d					
g	Strategic benefit		f	e			2	7	3	7.14
			g	f						
h	Characteristics of use		g				1	8	2	4.76
		h								
I	Sustainability						0	9	1	2.38
Sums								42	42	100

3. Each target criterion is then **assigned a measurand (unit)** (Table 6.3), which allows one to quantify the level of target achievement for each alternative. A currency (Euro) would be a reasonable unit for the profit. For properties that are difficult to quantify, such as the technical risk during implementation, a ranking of alternatives can be established, which is then used for evaluation.

4. Each alternative is now **assigned the respective property value**, and for each target criterion, respectively, the ranking is determined. Concerning the ranking, it must be carefully examined to what extent the achievement of the objectives of different alternatives is actually different. This applies, for example, to maximum requirements, such as "max. 15 kg component weight". If the list of requirements was carefully prepared, then all alternatives that meet this requirement must be seen as equivalent. On the other hand, relevant differences can be given for relative requirements (e.g. minimum weight). The evaluation of these differences leads to the question: Does the higher technical performance also increase the market value? Only then, there is a real advantage, which must be taken into account accordingly. Once again, the importance of a well-prepared list of requirements becomes clear. If the increase in technical performance beyond a minimum limit still brings an economic advantage, this must be stated accordingly. If the list of requirements only contains a certain target value, such as "max. 15 kg component weight", there is no reason why designs leading to 10 kg should be superior to those leading to 12 kg (besides material efficiency which is possibly a target criterion itself).

5. The property value is then converted into a **sub-target grade**, which should be between 0 and 10, with 10 being a complete achievement of objectives (to be compared with the list of requirements). Only through this transformation, does the weighted comparison of the criteria become meaningful, because the order of magnitude is unified. If a ranking of the alternatives has been defined, a corresponding sub-target grade can be defined by assigning the first-placed a value of 10, second-placed 9, etc.

6. Multiplying the weighting factors of the target criteria by the respective sub-target grades results in the respective **partial benefits**. For each alternative, these partial benefits are summed up to obtain the sum of benefit.

7. A decision can be made by **comparing the benefit values**. The alternative with the highest benefit value tends to contribute the most to maximizing benefits in a defined target system.

8. Finally, **sensitivity** should be investigated by examining the effects of varying weightings on the result. This way, an impression can be gained concerning the extent of the differences in the alternatives and the possible influence of subjectivity on the result. Overall, it is therefore advisable to perform the calculation using a spreadsheet program. This way, further target criteria can also be included without any problems.

In the example shown in Tables 6.2 and 6.3, alternative A would be preferable to alternative B. Although alternative B has a clear advantage, among other things, with regard to the strongly weighted achievable profit, the required investment, in

Table 6.3 Benefits analysis for overall techno-economical evaluation

Definition of target system			Alternative A			Alternative B		
Target criteria	Weighting	Measured value	Value	Sub-target grade	Partial benefit	Value	Sub-target grade	Partial benefit
Market price—costs	21.43	€	500	8	171.44	600	10	214.3
Invest costs	14.29	€	1 M	10	142.9	1.2 M	8.33	119.04
Development potential	11.90	Evaluation (scale 1–10[1])	4	4	47.6	6	6	71.4
Technical target achievement	9.52	Ranking[2]	2	9	85.68	1	10	95.2
Implementation risk	14.92	Evaluation (scale 1–10[3])	9	9	127.8	4	4	57.16
Development effort and duration	14.92	Evaluation (scale 1–10[4])	8	8	114.32	4	4	57.16
Strategic benefit	7.14	Ranking[5]	2	9	64.26	1	10	71.4
Characteristics of use	4.76	Ranking[6]	2	9	42.84	1	10	47.6
Sustainability	2.38	Ranking[7]	2	9	21.42	1	10	23.8
Sums	100	–	–	–	824.26	–	–	757.06

[1] 0 = high, [2,5,6,7] Lowest rank = most advantageous alternative, [3,4] 10 = low

combination with the high development costs and the associated implementation risk, more than outweighs this advantage. This is not untypical for product development with FRPs. A technically mature solution often competes with a quite novel and promising but less mature alternative. It is apparent that the definition of the preferences is of utmost importance, in order to make the right decision.

The sensitivity analysis described above can be used to check how a priority shift affects the result. The advantage of the benefit analysis is that it can be carried out relatively quickly. It also shows the extent to which alternatives differ, thus provides a good overview and helps to keep an eye on all relevant aspects. Given the possible relevance of the decision, however, it can be useful to use more complex decision-making tools, which offer a more reliable statement for the price while requiring higher effort. Corresponding methods can be found in [8] and [9].

With the preparation of the overall evaluation milestone 4 (Fig. 6.1) has been reached. Based on the result, a decision must be made whether the achievement of objectives is sufficient, or if further concepts should be elaborated and evaluated. This decision should be made by the entire development team and if necessary, with the involvement of the customer. If the decision is made to work out further concepts, the procedure is repeated from step 2–4 onwards (Fig. 2.1).

A basis for a decision is now available, which allows a final decision on the alternative to be realized. This decision is of course to be made by the customer and is the last step in the IPD methodology presented here (Milestone 5 in Fig. 6.1).

6.7   Questions for Self-Check

Below are some questions and tasks to help you reflect on the main contents of this section. The solutions can be found in Chap. 8.

> R44. An automotive supplier was entrusted with the production of a small series of engine bonnets. Based on the following data, calculate the costs per unit, using the method of process-based cost modeling. (component weight: 20 kg, material usage ratio for fibers and matrix: 125%, fiber weight ratio: 60%, costs for fiber material: 40 €/kg, costs for EP-matrix: 10 €/kg production by wet compression molding, fixed costs for production systems: 1.000.000 € per year (half of this is to be allocated to the bonnet production), energy costs: 50 € per component, personnel costs per component: 3 person hours, costs per person hour: 70 €, series size: 1000 pieces per year).
>
> R45. Name three cost categories that are assigned to the variable costs.
> R46. Name three cost categories that are assigned to the fixed costs.

R47. Name three strategic aspects that should be taken into account during the overall evaluation (besides the total manufacturing costs).
R48. Name three objectives that are pursued with the production of prototypes.

Literature

1. Bardmann, M.: Grundlagen der allgemeinen Betriebswirtschaftslehre. Gabler Verlag/Springer Fachmedien Wiesbaden GmbH, Wiesbaden (2011)
2. Corsten, H., Gössinger, R.: Produktionswirtschaft: Einführung in das industrielle Produktionsmanagement. Oldenbourg Verlag, Munich (2012)
3. Warnecke, H.-J., Bullinger, H.-J., Hichert, R.: Wirtschaftlichkeitsrechnung für Ingenieure, 2nd edn. Hanser Verlag, Munich (1991)
4. Schlimbach, J.: Ökonomische Prozessanalyse und Modellintegration zur Kostenberechnung von Faser-Kunststoff-Verbunden. IVW Publication series, vol. 64. Institut für Verbundwerkstoffe GmbH, TU Kaiserslautern (2006)
5. Hartmann, A.: Lebenszyklusberechnung als strategisches oder operatives Bewertungs-und Planungsinstrument für die Technologie der Faser-Kunststoffverbunde. IVW Publication series, vol. 11. Institut für Verbundwerkstoffe GmbH, TU Kaiserslautern (2000)
6. Holschuh, R. M.: Lokal lastgerecht verstärkte Multimaterialsysteme auf Basis von Polypropylen-Polypropylen-Hybriden. IVW Publication series, vol. 111. Institut für Verbundwerkstoffe GmbH, TU Kaiserslautern (2014)
7. Busch, J.V., Field, F.R., III.: Technical cost modeling. Hanser Verlag, Munich (1989)
8. Götze, U.: Investitionsrechnung: Modelle und Analysen zur Beurteilung von Investitionsvorhaben. Springer Gabler, Berlin Heidelberg (2014)
9. Poggensee, K.: Investitionsrechnung: Grundlagen–Aufgaben–Lösungen, 3rd edn. Springer Gabler, Berlin/Heidelberg (2009)
10. Haberfellner, R.: Systems enginsseering: Grundlagen und Anwendung. Orell Füssli (2012)
11. Saaty, T.L.: The Analytic Hierarchy Process: Planning, Priority Setting. Resource Allocation. RWS, Köln (1990)

Chapter 7
Closing Remarks

With the end of the presented methodology for integrated product development with FRPs, the preparation for the actual manufacturing of the product begins. Establishment of suitable supply chains, qualification of materials, development of processes and corresponding tools as well as many other activities are required, on the way to a fully functional serial production. Therefore, the importance of integrative measures exceeds the pure product development and extends over the entire product creation process. For a successful product creation, the specialists who are involved in all these upcoming phases must also develop a holistic way of thinking, as provided by this book.

The reader, to whom this book is intended to convey a generalist mind-set, might now also be at the beginning of further specialization in a specific field, e.g. component design, materials science or manufacturing. Like all other skills, the holistic way of thinking needs to be cultivated and trained, in order not to lose it. This is highly recommended especially to the students among the readers, who may still have many years of highly specialized training to go. Due to the rapid developments in the field of FRPs, a recurring consideration of all areas is essential. It must also be taken into account, that in the future, there will certainly be more specific and sophisticated methods for integrated product development, than the one presented in this book, which was idealized and simplified for teaching purposes. Hence, in addition to the different development areas for FRP, also the integrative measures require ongoing attention. Above all, the intrinsic motivation, to remain a generalist despite all specialization, needs to be preserved—because at the end, integration is only successful if it is truly lived by the engineers.

© The Author(s), under exclusive license to Springer Nature Switzerland AG 2021
D. May, *Integrated Product Development with Fiber-Reinforced Polymers*,
Engineering Materials, https://doi.org/10.1007/978-3-030-73407-7_7

Chapter 8
Response Catalog to the Questions for Self-Check

R1 Possible reasons for the usage of FRPs are, among others, the high lightweight potential, the resistance against environmental influences, the possibility to tailor the deformation behavior, the possibility for functional integration as well as the variety of materials and manufacturing processes.

R2 Division of tasks can become required when the technical complexity or the production quantity increases.

R3 Sequential and parallel division.

R4 IPD is the target-oriented combination of organizational, methodological and technical measures used by holistically thinking product developers [1].

R5 Human, methodology, technology, and organization.

R6 Interdisciplinary work between the disciplines of design, materials, manufacturing, calculation, and economy is essential. Four points are given as reasons:

- In principle, in addition to a geometrical design, a design to material and manufacture is required.
- The availability of reliable material parameters for dimensioning is problematic.
- Manufacturing restrictions determine the design freedom of the designer; neglecting this causes higher production effort and higher costs.
- The production technology determines the component quality and reproducibility.

D. May, *Integrated Product Development with Fiber-Reinforced Polymers*, Engineering Materials, https://doi.org/10.1007/978-3-030-73407-7_8

R7   Guidelines and scenario technique.

R8   Incomplete, inaccurate, incorrect, inadequate, and impossible.

R9   Design: basic functionality, weight targets, part geometry;
      Manufacturing: intended series size, geometric tolerances, specifica-
      tions with regard to joining methods for target component with a
      superordinate structure.

R10   Wishes can be distinguished in explicit and implicit. Explicit wishes
      are directly stated by the customer. Implicit wishes are not stated, e.g.
      because they seem obvious to the customer or the customer isn't even
      aware of them.

R11   Fixed demands define a mandatory characteristic of a product. Interval
      demands define a target range (lower and upper limit) for a specific,
      quantifiable value.

R12   False.

R13   True.

R14   True.

R15   False.

R16   True.

R17   Amorphous thermoplastics: very long, randomly oriented and entan-
      gled molecular chains; Semi-crystalline thermoplastics: Long molec-
      ular chains that are partially regularly arranged; Thermosets: firmly
      linked, three-dimensional molecular networks.

R18   The fibers attract the forces due to their comparatively high stiffness.
      The matrix embeds the fibers, fixates them, supports them under
      pressure, thus enabling the load to be introduced into the fibers, pro-
      vides load distribution between the fibers and protects the fibers from
      environmental influences. The interphase determines the adhesion
      between fiber and matrix.

R19   True.

R20   False.

R21   Influence of the fiber length see Fig. 4.4.

R22   Influence of the fiber orientation see Fig. 4.26.

R23   The terms can be defined as follows:

- **Material lightweight design**: Measures to minimize weight based
  on usage of materials with relatively high specific properties (such
  as FRPs).
- **Structural lightweight design**: Adaptation of the structure to the
  given load situation and resulting load paths.
- **System lightweight design**: Multifunctionality of components
  through functional integration beyond the load-bearing function
  (e.g. load-bearing lightning protection).

R24 Advantages of a differential design are, e.g., the reduced complexity of the single manufacturing processes, the improved possibilities for replacement of single parts in case of damage as well as the simplicity concerning combination of different materials. On the other hand, integral design allows a relatively homogeneous flow of forces, the need for only one manufacturing process (ideally) and the quite small assembly effort.

R25 Types of deformation couplings are: strain-bending, strain-shear, strain-torsion, shear-torsion and bending torsion. A $[0°, 90°, 45°]$—layup would show all of these couplings.

R26 From the FLEA-equation $\left(u_x = \frac{F_{max} \cdot L_1}{E_x \cdot A} = \frac{4000N \cdot 1000mm}{E_x \cdot 50mm^2} < 2mm \right)$ one can derive that the stiffness must be at least 40 GPa. The minimum fiber volume content of 56% can be calculated via the rule of mixtures for the young's modulus (parallel connection of springs).

R27 Thermoset matrix systems provide a lower processing viscosity, as the polymerization takes place directly in the manufacturing process. They are well-suited for adhesive bonding and show high temperature-stability. Yet, shelf life is limited, the toughness is quite low, and curing is irreversible.

R28 Thermoplastic matrix polymers can be re-molten and provide a relatively high toughness. Furthermore, semi-finished products based on thermoplastics have a virtually infinite shelf life. Yet, the processing viscosity is quite high, thermoplastics tend to creep and adhesive bonding is challenging (especially for semi-crystalline thermoplastics).

R29 The deformation behavior of semi-crystalline thermoplastics is illustrated in Fig. 4.55 (bottom).

R30 Impregnation describes the process of wetting the individual filaments and filling spaces between the filaments with liquid matrix. During consolidation, the target fiber volume content is set by pressure, which also counteracts restoring forces. If air is present in the reinforcement structure, it is driven out and formation of new air inclusions is prevented. If the FRP is a laminate, the individual layers are bonded to each other. Solidification describes the transition of the liquid matrix into solid state, by either solidification (thermoplastics) or chemical cross-linking (thermosets).

R31 Advantages of FRPs are, among others, the very good specific properties, the variety of materials and manufacturing processes, the design possibilities, the corrosion resistance, and the very good possibilities for functional integration. Disadvantages are the often high manufacturing costs, especially in high-performance applications, the complexity of repair and recycling measures, the deficits concerning standardized design methods as well as the challenging mechanical processing and joining.

R32 The fiber paradox can be explained, above all, by the size effect, molecular alignment as well as the defect orientation.

R33 False.

R34 False.

R35 True.

R36 False.

R37 When choosing a matrix polymer the limits concerning the application, the chemical resistance, the appearance as well as the costs should all be considered.

R38 Joining FRPs can be challenging, e.g. due to contact corrosion, differing coefficients of thermal expansion, creep as well as FRP-suitable load introduction.

R39 Damage analysis of FRP parts can be carried out, e.g. by ultrasonic scanning, thermography and shearography.

R40 During recycling of continuous fiber-reinforced polymers, the fibers are shortened. During further usage, the reduced length and the thereby complicated orientation of the fibers can cause deficiencies concerning the exploitation of the mechanical potential.

R41 For systems used in the production of FRPs, among others, the investment costs, the temperature ranges, the movement velocities and the press forces are relevant.

R42 Heat can be introduced by conduction (e.g. direct contact of SMC compound with a hot tool), convection (e.g. heating an organo sheet in a fan oven) and/or radiation (e.g. heating an organo sheet in an infrared heating field).

R43 1f, 2a, 3b, 4d, 5g, 6c, 7e.

R44 Given a fiber weight fraction of 60%, the 20 kg part contains 12 kg fibers and 8 kg matrix. With a material usage ratio of 125% (for fibers and matrix), this means that per part, 15 kg of fibers and 10 kg of matrix polymer are required. Multiplied with the material costs, this sums up to 700 €. The labor costs per part are 3 h * 70 €/h = 210 € and the energy costs are 50 €. Hence, in sum the variable costs are 700 + 210 + 50 € = 960 €. Only half of the investment costs of

1.000.000 € is to be accounted to the considered product and is divided by 1000 parts per year. The fix costs are therefore 1.000.000/ (2 * 1000) = 500 €. In sum, the costs per part are 960 + 500 € = 1460 €.

R45 The variable costs comprise, e.g. the energy costs, the labor costs and the material costs resulting from the product manufacturing.

R46 The fixed costs comprise, e.g. the depreciation, the space costs and the interests.

R47 For the overall evaluation of a draft the development time, effort and risk as well as the strategic benefit should be considered.

R48 By building prototypes, the technical performance can be proven, the suitability of selected semi-finished products can be verified and the material suitability can be validated.

Literature

1. Ehrlenspiel, K., Meerkamm, H.: Integrierte Produktentwicklung: Denkabläufe, Methodeneinsatz Zusammenarbeit. Carl Hanser Verlag GmbH Co KG, Munich (2013)

Index

Printed in the United States
by Baker & Taylor Publisher Services